Cyberchiefs

Cyberchiefs

Autonomy and Authority in Online Tribes

MATHIEU O'NEIL

PLUTO PRESS
www.plutobooks.com

First published 2009 by Pluto Press
345 Archway Road, London N6 5AA and
175 Fifth Avenue, New York, NY 10010

www.plutobooks.com

Distributed in the United States of America exclusively by
Palgrave Macmillan, a division of St. Martin's Press LLC,
175 Fifth Avenue, New York, NY 10010

British Library Cataloguing in Publication Data
A catalogue record for this book is available from the British Library

ISBN   978 0 7453 2797 6   Hardback
ISBN   978 0 7453 2796 9   Paperback

Library of Congress Cataloging in Publication Data applied for

10 9 8 7 6 5 4 3 2 1

Designed and produced for Pluto Press by
Chase Publishing Services Ltd, Sidmouth, England
Typeset from disk by Stanford DTP Services, Northampton, England
Printed and bound in the European Union by
CPI Antony Rowe, Chippenham and Eastbourne

For GHM

CONTENTS

ACKNOWLEDGEMENTS

Elements of Chapters 1 and 5 appear in a different form in Mathieu O'Neil, 'Radical Tribes at Warre: Primitivists on the Net', in Tyrone L. Adams and Stephen A. Smith (eds), *Electronic Tribes: The Virtual Worlds of Geeks, Gamers, Shamans and Scammers*, Austin: University of Texas Press, 2008.

Some ideas developed in this book were first presented at the following events: Blogtalk Downunder (University of Technology Sydney, 2005); Workshop on Internet Mediated Sociality (Academy of the Social Sciences in Australia, 2006); Social and Political Theory seminar (Research School of Social Sciences, Australian National University, 2007); Australian Demographic and Social Research Institute seminar (Australian National University, 2007). Thanks to the participants for their feedback.

I owe the term 'index authority' to Rob Ackland of the Virtual Observatory for the Study of Online Networks. Thanks are also due to the following for their input: Vidisha Carmody, Lincoln Dahlberg, Adrian Hayes, Seth Keen, Ted Mitew, Tim Phillips, James Rice, Jennifer Rutherford, Rob Schaap, Russell Smith, Jodie Vaile, Judy Wajcman, Wan-ling Wee, Michael Wood. Thanks to the anonymous reviewers for their constructive criticism of my proposal. Special thanks to Kaspar Møller Hansen and Icon Tada. Eternal gratitude to Charmian and Mike. All my love to Hari, Mira, and Gita. Big Up to Claire, Marc-Antoine, Sumitra, Ramanand, Vasanth, Nicolas, Alex, Frank, Marco, Christel, Silvain, Philippe, Yohannes 'B-C', Serge, Lucy, Alvin, Yvonne, Vincent, Gilles.

INTRODUCTION

C'est en ce sens qu'il est permis de penser que la vérité de ce rapport sur mon temps sera bien assez prouvée par son style. Le ton de ce discours sera en lui-même une garantie suffisante, puisque tout le monde comprendra que c'est uniquement en ayant vécu comme cela que l'on peut avoir la maîtrise de cette sorte d'exposé.*

Guy Debord, *Panégyrique*

The defining fact about the Internet is that it is a network, a collection of nodes connected by ties. Any node on the Internet is accessible from any other node, and there are no differences between the ties that connect the nodes: all hyperlinks are equal. In liberal democracies, this many-to-many structure and the informality of online social relations are taken to mean that cyberspace allows people to freely engage in social and political exchanges with others who share common interests. Though inequalities of access persist, goes this argument, the Internet has become a prime avenue for spontaneous expression and organisation.[1] Online sociality is said to reject hierarchy, creating a sort of permanent autonomous zone of democratic communication and production. In the realms of independently produced media, knowledge and code, participatory cooperation is the rule. Pyramidal structures and proprietary practices are being inexorably challenged by a swarming multitude of self-organised agents.

In reality, *authority* runs rife on the Internet. Online self-organisation and self-expression, in order to avoid an incoherent Babel, require participants to exercise quality control over their work and the membership of their groups. Participants need to be able to determine who is reliable; what contributions are pertinent;

---

* It is in this sense that the truth of this report on my time will be well enough proved by its style. The tone of this discourse will in itself constitute sufficient guarantee, for everyone will understand that it is only by dint of having lived in this way that one can master this kind of account.

and, on that basis, who will be included or excluded, reinforcing the feeling of belonging. This book focuses on authority in online projects or 'tribes'. Since the Internet is a stateless system, the interactions which occur on it can properly be called 'tribal'. Online tribes are social formations which favour grassroots direct democracy, the pleasurable provision of free gifts, and the feeling of proximity to others. Max Weber classically defined authority as the recognition by others of a person's *legitimate right* to exercise power. The question this book addresses is: How does authority take into account the central value on the Internet, *autonomy*? Analysing authority necessitates an interrogation of the notions of expertise and leadership; ultimately, it raises the question of the nature of domination.

Michel Foucault suggested that expertise is an instrument of elite domination: in his view, the state used scientific experts to define individuals and groups as deviant or sick, and to justify their discriminatory treatment. But specialised knowledge has also been used for autonomous purposes. Computer engineers or 'hackers' (not to be confused with computer vandals or thieves) created the Internet. If computer code was efficient and elegant, if it worked, its author was rewarded with high status. Quasi-scientific expertise became independent from hierarchical institutions: hackers recognised the judgment only of their peers. The authority of *experts* is traditionally subordinated to the authority of *leaders*. However when the Internet was developed *learned authority* to a great extent determined *administrative authority* for the simple reason that only computer hackers knew how to run the systems. Following the lead of hackers, expertise on the Internet became dependent not on credentials issued by an institution to an individual, in the shape of a diploma or professional certificate, but on an individual's unique skill, developed over time, and publicly demonstrated. In his memoirs, Guy Debord declared that only someone who had lived a life apart from, and against, capitalism and its media propaganda (the 'society of the spectacle') could write in classical French: the truth of his account would be proved *by its style*. Online communication similarly requires public performances blending humour, profanity and knowledge to confirm that expertise is both *authentic* and valid.

Beyond communication, the user-generated social Internet ('Web 2.0') is increasingly a site of peer production, of cooperative work. Distributed projects involve thousands of people, located in different places and submitting at different times contributions that vary widely in scale. How are these contributions assembled? More broadly, how does domination operate in self-directed networks? And how should we go about finding out? To answer these questions, we have first to recognise that, on the Internet as everywhere, the playing field is never level. Structural inequalities exist because of the effects generated by the Internet's growth pattern, which privileges early entrants. In addition, archaic forms of power, such as overt sexism, are rife online. Pierre Bourdieu once said that sociology was the 'science of domination'. Bourdieu's critical approach provides invaluable tools for understanding the reproduction of privilege. But does this mean that only *sociologists* can understand the truth of power? In other words, is everyone else a 'cultural dope'? Not in the least: archaic power can be contested. Moreover, people are capable of making judgments about what is at stake in conflicts, what the roles of leaders should be, and whether decisions are fair. The analysis of online domination must take these different dimensions into account.

The distribution of administrative authority to autonomous individuals is an essential part of the appeal of volunteer projects such as Wikipedia, precisely because it gives people the possibility of rapidly attaining positions of authority. In the online context, administrative authority is the capacity to exclude people from a network, or to limit the actions they can undertake on that network. To understand how autonomous social formations justify these actions, it is useful to look at David Beetham's contention that two legitimising principles are more emancipatory than others. The first is the principle of democratic *sovereignty*, based on the collective will of the group. The second is the *meritocratic* principle of differentiation which, in theory, challenges the reproduction of advantage. On the Internet, meritocracy was separated by hackers from hierarchy and bureaucracy. Merit assumed an anti-authoritarian slant, based on the regard for the charismatic genius of great initiators, and, subsequently, on the charismatic position of great nodes. The principles of autonomous charisma and democratic sovereignty structure the online space of authority.

The Internet has proved highly popular with researchers of all stripes because it provides free and easy access to innumerable traces of human and network activity, whether in the form of text or hyperlinks. Scores of empirical studies, both quantitative and qualitative, have been carried out. Less common have been comprehensive conceptualisations of issues other than virtual identity and community. Why is it necessary to create new frameworks for the analysis of online authority? Because the primary aim of domination is to be misrecognised, and what better misrecognition could there be than the widespread notion that the Internet is a *non-hierarchical* space? And yet: the persistence of some forms of domination should not prevent us from recognising instances where authority really is self-directed.

Internet research, since it is still an emerging field, offers a welcome opportunity to break through disciplinary silos. Though this book's main thrust is sociological, its conceptual toolkit draws from communication and new media studies, anthropology, political theory, network theory and law. The first part of this book reviews concepts useful for the analysis of online sociality. Chapter 1 focuses on the role of autonomy within informational capitalism and on the emergence of neo-tribalism. Chapter 2 examines the impact of anti-authoritarian meritocracy, distribution and aggregation on online charisma. Chapter 3 considers structural determinations such as network growth and archaic sexism. Chapter 4 examines justification, governance and law in online tribes. This first part concludes by presenting a model of the space of online tribal authority, structured along two main axes: charismatic and sovereign authority. Part II examines four projects representing distinct locations in the space of online authority.[2] In each case the focus is on the key characteristics of the project, its authority structure, and the nature of the conflicts it generates. Chapter 5 looks at the

Primitivism radical text archive. Chapter 6 analyses *Daily Kos*, a progressive political community weblog. Chapter 7 focuses on the *Debian* free-software project, particularly mailing lists. Chapter 8 examines *Wikipedia*, the encyclopedia that anyone can edit. Chapter 9 pulls the book's strands together and argues that online authority should be understood in the context not of networks, but of a new organisational form, *online tribal bureaucracy*.

PART I

1

THE AUTONOMY IMPERATIVE

Are we who live in the present doomed never to experience autonomy, never to stand for one moment on a bit of land ruled only by freedom?

Hakim Bey, *The Temporary Autonomous Zone*

Autonomy refers to people's capacity to be authors of their own lives, to exercise self-determination and self-government. Why has autonomy become a central concern in contemporary society? Manuel Castells describes our age as dominated by *media politics*. The means to access state power is through the mobilisation of votes, and since people form their political opinion through the mass media, 'outside the sphere of the media there is only political marginality'.[1] As a consequence, professional politics is dominated by features which accord with the media format, such as personalisation, image-making, financial dependence on interested donors and scandal politics. Media politics leads to a loss of trust by the public in the political process, and to a crisis of credibility of the political system.[2] This crisis is compounded by globalisation. Liberal democratic citizens who are concerned about global social and environmental issues find it difficult to address them through normal political channels, such as local or national elections, partly because of the loss of legitimacy of political parties, and partly because national politics has limited impact over global issues. A consequence of this crisis of collective identities and institutions is that political engagement is, increasingly, self-constructed, and that there is growing interest in the possibility of exploring political activity outside the traditional party system.

As the rise of the Internet occurred at roughly the same time as liberal democratic institutions were coming to be perceived as experiencing a legitimacy crisis, these two phenomena have been connected in an optimistic way by numerous commentators.[3] The Internet does offer opportunities to build up self-directed networks of horizontal communication at little cost, bypassing institutional controls. This is said to represent a new kind of *networked politics* which allow the possibility of establishing multidirectional connections to many individuals.[4] In so doing, people are reconnecting with the age-old aspiration of living autonomously. This chapter provides an overview of online autonomy, discusses its relationship to informational capitalism and assesses its suitability as a conceptual tool for the analysis of Internet sociality.

Anarchism and the Global Network of Struggles

By stressing the importance of autonomy, Internet research is harking back to a long tradition of radical politics based on *self-organisation*, understood as 'a process of order formation that comes from within a system'.[5] The preoccupation with grassroots politics and the questioning of formal leadership structures characteristic of networked autonomists derives from previous generations of dissenters, starting with nineteenth-century anarchists. For Kropotkin, human activity should be founded on cooperation rather than competition, whilst Proudhon and Bakunin argued that society should be a collection of autonomous individuals living freely in a federation of communities: 'We want the reconstruction of society and the unification of mankind to be achieved, not from above downward by any sort of authority, nor by socialist officials, engineers, and other accredited men of learning – but from below upwards.'[6] These founding principles are still advocated today by many anarchists.[7] A good example of a successful network of self-managed companies is the Mondràgon Cooperative Corporation which has operated in Spain since the 1950s.[8]

Beyond the strictly anarchist sphere, the twentieth century saw a fertile cross-pollination between anti-authoritarian ideals and other forms of radical critique, chief amongst which was Marxism.

Examples include Amadeo Bordiga's critique of Stalinism as 'state capitalism' and Karl Korsch and Anton Pannekoek's advocacy of giving all power to workers' councils. These thinkers in turn influenced 1960s 'New Left' French radical groups such as *Socialisme ou Barbarie* and the *Internationale Situationniste*. These groups criticised both Left and Right totalitarianism, rejecting all power hierarchies and in particular the institutional pressure of the Communist Party and its bureaucracy.[9] They pointed to the basic paradox of traditional socialism, which purports to fight inequality through increased rationalisation and a hierarchy of cadres and specialists, thereby resulting in less freedom.

During the same period Italian thinkers gathered around the figures of Toni Negri and Mario Tronti created *Potere Operaio*, an influential group and magazine which dissolved into the *Autonomia* movement in 1973, and whose core principle was 'autonomy at the base', the belief that workers can force change in the capitalist system by themselves, without the mediation of political professionals. For Italian autonomists the working class included people involved in non-unionised work, such as housework and study.[10] Other examples of radical autonomy include Murray Bookchin's ecological anarchism, where a 'commune of communes' or confederation of face-to-face assemblies allows people to directly manage society,[11] and Cornelius Castoriadis's definition of democracy as a permanent bottom-up process, identical to autonomy or self-institution.[12]

In general New Left activists espoused anti-authoritarianism, direct action, decentralised organisational forms and the combination of culture and politics for the transformation of everyday life, all of which were apparent during the student revolts of 1968. Ultra-left movements which emerged in Europe in the 1970s and 1980s (such as the German *Autonomen*) adhered to the principles of collectivism, self-determination and decentralised direct democracy which were expressed in practices such as consensus-based decision making, conscious spontaneity, militancy and confrontation as tactics.[13] The punk rock movement which began in the late 1970s and has continued to exist in various guises to this day also played a role in fostering the theme of self-determination. Though the thousands of do-it-yourself (DIY) projects which originated from it were often limited to the cultural sphere of the production of records, tapes, fanzines, and concerts, the punk 'scene' afforded participants the opportunity to establish or take part in squats, social centres, food cooperatives (such as *Food Not Bombs*) and actions (such as the *Critical Mass* protest bike ride).

Examples of autonomous projects exist all over the world. In Italy for example, more than 100 squatted *Centros Sociales* (Social Centres) have served as hubs for political organising, autonomous social services and radical culture.[14] Defending threatened social centres motivated the creation of radical Italian groups such as the *Tute Bianche* (White Overalls) and *Disobbedienti* (Disobedients) and was central to their protest actions. In Italy, as elsewhere, such spaces and projects parasitically exist on the fringes of the dominant system: punks and squatters still depend, to various degrees, on states and corporations for food, energy, healthcare and security.

Full collective autonomy is therefore only possible when a group controls all these basic necessities and uses them for its own purposes, such as when a dispossessed community rises up against its oppressors. This type of revolt was affected, during the 1990s, by globalisation: the delocalisation of work, the technologically enabled instantaneous movement of capital, and the pressure to deregulate state welfare programmes. Transnational corporations and the international organisations which regulate the world market (such as the International Monetary Fund, the World Bank and the World Trade Organisation) became the symbols of this process. In neoliberal structural adjustment programmes, collective grassroots movements of peasants and workers found both a justification for a place-centred desire for self-rule, as well as opportunities to connect globally to other under-represented groups via new communication technologies.[15] An archetypal struggle is that of indigenous groups protecting their homeland from industrial development, such as the U'wa in Colombia and the Ogoni in Nigeria. By far the most influential indigenous group has been the Zapatista Army of National Liberation (*Ejército*

Zapatista de Liberación Nacional, EZLN), which rose up against the Mexican state and its neoliberal policies on 1 January 1994, the day the North American Free Trade Agreement (NAFTA) came into being. Two days later the EZLN spokesman, the balaclava-wearing Subcomandante Marcos, started publishing his poetic declarations from the Lacandona forest online, inaugurating the age of the electronic social movement. The Zapatistas provided a crucial impetus to the global networking of local autonomous groups opposing neoliberal economic policies.

The network form allowed Global Justice groups to collaborate whilst maintaining their autonomy. The People's Global Action (PGA) network was created during the first Zapatista 'International Encounter for Humanity and Against Neoliberalism' (*Encuentros*), held in the small town of La Realidad in Chiapas, Mexico. The EZLN's Marcos declared on 3 August 1996 that the PGA would disseminate and renew the Zapatistas' rebel voice, as 'an echo that turns itself into many voices, into a network that before Power's deafness opts to speak to itself, knowing itself to be one and many', leading to 'a multiplication of resistances'.[16] The Zapatistas crystallised the attention of sympathisers and supporters worldwide, leading to the establishment of a global network of solidarity committees, such as *Ya Basta!* in Italy, and sparking mention of a 'Zapatista Effect'.[17]

The alliance of new social movements (such as the peace, women's and ecological movements), progressive trade unions and oppositional youth cultures which became known as the Anti-Globalisation or Global Justice movement was inspired by these tactics. They used new communication technologies to coordinate other campaigns (such as that against the Multilateral Agreement on Investment or MAI in 1998), protest actions (such as the 18 June 1999 'Carnival against Capital!', the 'Battle of Seattle', which shut down the WTO meeting in December 1999 and subsequent anti-WTO demonstrations), and meetings (such as other *Encuentros* and the World Social Forums). The crucial importance of the Zapatistas in connecting diverse strands of global protest was shown by a study of the global network of activist non-governmental organisations (NGO) and grassroots groups in 2001: without the pro-Zapatista cluster at its centre, binding the network together, this set of websites would have been much more balkanised.[18]

The Global Justice movement's embrace of the Internet derives from the New Left's anti-hierarchical strain. It has become common to think not of a single movement, but of a 'movement of movements', in which a fluid constellation of groups is difficult to control, monitor and police. This organisational structure is intended to guard against the formation of hierarchies and the centralisation of power: no central committee distributes the correct 'line' of resistance. The absence of fixed points or centres means that themes are created and disseminated through multiple networks and connections, formed and maintained by forums and gatherings.[19]

Castells argues that networks are, theoretically and practically, perfectly suited to autonomous expression: the contradictory nature of such movements is a source of strength rather than weakness. Divisions and differences add new support because many individuals recognise themselves in at least one of the facets of the movement, and do not feel themselves subject to the pressure or discipline of those factions with which they disagree. Dysfunctional nodes that block the overall dynamic of the network can be switched off or bypassed, thus overcoming the traditional ailment of social movements so often engaged in self-destruction through factionalism.[20]

Self-realisation in the Digital Commons

The history outlined above illustrates an important dimension of autonomy online but does not fully account for the practices of people in Web 2.0, where a privileged part of the population sees the Internet as the means for self-realisation. Though hackers, bloggers and wikipedians may sympathise with social movement activists, and occasionally support them, they cannot be wholly identified as activists. The primary purpose of social movements is a struggle for justice by means of establishing and maintaining alliances with a variety of institutional actors and other social movement organisations. The energies of bloggers, hackers or wikipedians are principally directed towards building coherent autonomous systems, such as independent media, encyclopedias or software suites. Rather than fitting within the categories of activism, where the Internet offers the means of advancing an offline cause, *online practice in and of itself* is perceived as a worthwhile experience.

Online autonomy first manifested itself in the realm of the open sharing of information for code production. The US Defense Department's Advanced Research Projects Agency established the ARPANET network in September 1969. As is well known, this forerunner of the Internet owed its decentralised form to the necessity of withstanding a nuclear attack. Data flows would be able to arc around any shattered nodes. The North American computer engineers and particularly the graduate student hackers who invented distributed network technology did not object to being funded by the Pentagon. Yet these programmers were infused with the values of individual freedom, of independent thinking and of sharing and cooperating with their peers, values which characterised 1960s student culture. This culture was translated into open technical standards and the belief that individuals are being liberated to the extent that they can now work for pleasure, to satisfy their natural curiosity and because of the appreciation they receive from their peers, following a 'hacker ethic' of sharing and cooperation.[21]

Historically, the development of free and open-source software (FOSS) was dependent on the mass advent of the Internet as a software delivery system permitting previously undreamed of economies of scale, such as practically zero-cost distribution, and instant collaboration across national boundaries of highly competitive computer *aficionados*. Collectively, they produced software programs for web servers (Apache), email (sendmail) and database management (Perl) which allow the Internet to run, whilst the GNU/Linux operating system has posed a robust challenge to Microsoft's Windows, the dominant proprietary system.

Yochai Benkler has argued that free software and other peer-produced projects such as Wikipedia depend on the maximum autonomy of participants. Peer production represents a real alternative to the dominant production models organised around commands and hierarchies (as in firms) and prices and monetary rewards (as in markets).[22] Decreased communication costs and the fact that digital goods are non-rival (one person's use does not hinder another's) contribute to making peer production an attractive alternative to markets and firms, matching best available human resources to the best available information inputs to create information products.[23] If huge numbers of people contribute, all the better: people are best able to decide themselves how much they can contribute. It is true that people may mistake or misstate their capacities, but peer review or statistical averaging (if the number is large enough) will be enough to control bad self-assessments.[24]

Projects will succeed if they are *modular*, signifying they can be broken up into distinct components which can be independently developed, allowing investments at different times of distinct individuals with varying competencies. Projects should also be *granular* (modules need to be fine-grained) so that they can be performed by individuals in little time, and motivation needs to be very small.[25] The modules should be of different sizes to accommodate heterogeneous motivation levels.[26] Given the relatively small value such fine-grained contributions will have and the high cost of remunerating each contribution monetarily, non-financial rewards, such as the pleasure of creation, will be more effective in motivating large-scale peer-production efforts.[27]

Apart from peer production, the development of free software was also made possible by a set of legal licences which give users the right to freely access, modify and redistribute the source code of any software program protected by the licence; it also compels them to provide these rights to anyone else. The most well-known of these licences is the General Public License (GPL), or 'copyleft'. According to Richard Stallman, creator of the GNU system, of the GPL and of the Free Software Foundation, his overall purpose was to extend as far as possible the boundaries of what could be done with entirely free software: 'The idea of GNU is to make it possible for people to do things with their computers without accepting [the] domination of somebody else.'[28]

A derived set of autonomous ventures are peer-to-peer distribution networks, which have been used for the exchange of free or 'pirated' data by enthusiasts, as well as the rise of peer-production projects such as open journalism and wikis. Autonomous Net journalists aim to contest the legitimacy of the corporate media, by establishing a space free from the power elites which control mainstream journalism. Today, direct communication without professional mediation is what defines the 'blogosphere'. But before weblogs, this was also true in the case of the decentralised transmission, unregulated expression, self-legislation and local control over decisions characteristic of the worldwide network of Independent Media Centers (IMCs). The Indymedia model was based on the free-software model of sharing, creating and interacting at no cost and for the benefit of all.[29]

Wikis (the Hawaiian word for 'quick') are hypertextual archives based on open editing. The Wikipedia online encyclopedia, which was launched in January 2001 is the most famous example. Any online user can change not only the content of a Wikipedia article (by adding, editing, or deleting material) but also the site's organisation (by creating links, for example). In addition, wikis contain a built-in fail-safe mechanism which automatically records all modifications. Users are then able to transform the archive as they see fit, as no version of the previous information is ever irredeemably lost. Another example is Project Gutenberg, which aims to create a globally accessible library of public-domain e-texts.

Theories of Online Autonomy

The inspiration for Castells's concept of the 'network society' is the branch of economic sociology known as social network analysis. The Canadian sociologist Barry Wellman wrote that 'when a computer network connects people it is a social network'.[30] Social network analysis considers social relations as arrangements of individual *nodes* organised in *clusters* and connected by *ties*. The structure of relations among actors and the location of actors on networks is held to have important behavioural, perceptual and attitudinal consequences, both for the individual units and for the system as a whole. Social action is understood in terms of structural constraints on activity and opportunities for gaining advantage, rather than assuming that inner forces (such as internalised norms) impel actors towards goals. Social network analysis provides a series of metrics to analyse interactions: the centrality of actors, the density of networks and the formations of clusters or blocks can be precisely measured. An important concern is how networks allocate flows of scarce resources to system members.[31] It is clear that social network analysis offers key concepts and methodologies for the study of online networks, and that its focus on visualisation enables the quick communication of complex social relations. However, social network analysis is ill-equipped to address the broader significance of the proliferation of autonomous activity on the Internet, because it eschews any discussion of culture and ideology. By 'ideology' I do not mean an illusory moralising discourse intended to conceal material interests, but, following anthropologist Louis Dumont, rather a set of shared beliefs, inscribed in institutions and bound up with actions.[32]

Since utopian political solutions are no longer considered likely to occur offline, the Internet has come to embody the spirit of Utopia. In such a charmed universe everyone can have a say, from 'cyberlibertarians' who decry the influence of governments to 'cybercommunists' who believe that peer production will revolutionise both the market economy and traditional hierarchy. The primary tenet of the ideology of the Internet is that *online networks are privileged sites for the flowering of freedom*. Following its inception in the hacker universe, this ideology was disseminated by writers and activists who celebrated the Internet's potential for empowerment and 'resistance' to power. The Electronic Frontier Foundation (EFF) was established by Mitch Kapor and John Perry Barlow in 1990 to 'help civilise the electronic frontier' in keeping with 'society's highest traditions of the free and open flow of information and communication'.[33] In 1996 one of the EFF's founders, the noted cyberlibertarian John Perry Barlow, wrote *A Declaration of Independence of Cyberspace* in which he advised the 'Governments of the Industrial World' to 'leave us alone. You are not welcome among us. You have no sovereignty where we gather.'[34] A technology pundit asserted in the pages of *Wired* magazine that 'being digital' could flatten organisations, globalise society, decentralise control and help harmonise people.[35] A slightly different take was put forward by cultural critics such as Sherry Turkle and Mark Poster. In their view, many-to-many communication, with the simultaneous reception, alteration and redistribution of cultural objects, frees the subject from the territorialised relations of modernity; the Internet is the material expression of the philosophy of postmodernism.[36]

Another major component of the Internet ideology is that *online networks subvert capitalism*. The hypermedia theorist Richard Barbrook wrote that the invention of the Internet was the greatest irony of the Cold War as, at the height of the struggle against Stalinist Communism, the US military unwittingly financed the creation of 'cyber-communism'.[37] As information is incessantly reproduced, the quantity of collective labour embodied in each copy is soon reduced to almost nothing, and the Internet's very structure threatens the dominant order. In Barbrook's view, Americans are enthusiastically 'superseding' capitalism by practising 'really existing anarcho-communism'.[38] In short, the Internet exemplifies a new, subversive model of organisation, a high-technology gift culture which contradicts techno-capitalism's 'Californian Ideology', understood as the mix of the freewheeling spirit of hippies and of the entrepreneurial zeal of the yuppies.[39] A similar point is made by McKenzie Wark when he suggests that the activities of what he calls the 'hacker class', being based on the free manipulation and exchange of inexhaustible digital information, challenge the very basis of the process of accumulation: new hacks supersede old hacks, and devalue them as property.[40]

The notion that online networks are anti-capitalist also appears in Toni Negri and Michael Hardt's depiction of the Internet as a terrain for what they call the 'multitude', the irreducible singularities of autonomous subjects (in preference to unitary 'people') who produce objects, services and knowledge, but whose labour is dispossessed by capital. For such thinkers, operating in the Italian *Autonomia* tradition, autonomy is the independence of social time from the temporality of capitalism. In concrete terms, this means a refusal or retreat from capitalist social relations – the strategy of the refusal of work, strikes and sabotage. But when power itself becomes networked and distributed, and when everyone is interconnected, where does one escape to? Negri suggests that people can withdraw to the Internet, particularly the non-corporate areas.[41] In Hardt and Negri's view, the Internet, a prime example of a completely horizontal and deterritorialised democratic network, is what Deleuze and Guattari called a 'rhizome'.[42] The rhizome, originally a subterranean process of plant growth involving propagation through the horizontal development of the plant stem, has become the canonical expression of a non-hierarchical, decentralised network in which any point is connected to any other point.[43] Marx's single old mole of the proletariat boring through the factory's floor becomes a 'tribe of moles' digging interconnected tunnels.[44] Neo-*Autonomia* theorists describe the liberatory potential of a collective intelligence autonomously produced by immaterial labourers such as hackers.[45] The democracy of the multitude is as an 'open-source society', a society in which source code is revealed so that 'we can all work collaboratively to solve its bugs and create new, better social programmes'.[46]

The Internet Ideology and Informational Capitalism

Theories ascribing a revolutionary potential to the Internet raise a number of questions. How can peer production and the Internet ideology of freedom threaten a type of market relations defined by Castells as 'informational capitalism', in which the generation, processing and transmission of information are the fundamental source of productivity and power?[47] Are not gifts and capital ultimately irreconcilable, and is the market economy not always threatening to reprivatise the common enclaves of the gift economy?[48] Matthew Hindman notes that the development of open-source software was 'in large part the result of competition within the technology industry'.[49] Major corporations such as Sun, Motorola, Apple, and Oracle have supported open-source development to attract contributors so as to undermine Microsoft's market share. In 2000, Tiziana Terranova argued that in addition to modifying software packages, the provision of immaterial 'free labour' such as building websites, reading and participating in mailing lists, building virtual spaces on multi-user dungeons (MUDs) is a fundamental moment in the creation of value in capitalist digital economies.[50] What might have been seen as a controversial assertion at the time has become blindingly obvious with the mass rise of Web 2.0 and, in particular, of social networking platforms: consumers are now *themselves* expected to provide the content which will then be used to attract advertising revenue, so that 'cooperation is used for advancing the logic of capital accumulation'.[51]

The diffusion of free software allows its authors to reach people who would otherwise refuse to purchase it; and if others decide to diffuse it also, they will always appear as less competent in offering software-related services (such as code maintenance, user guides and updates) than the originators: the authors' competitive advantage is seldom threatened. Furthermore, peer production provides a response to the contemporary quest for authentic, personalised, community-type advice provision. Contemporary consumers increasingly consider markets as intermediary stages where standardised products will be assembled and configured according to their needs. Such bricolage was pioneered by computer enthusiasts. Unlike driving a car or working with tools, using a computer was (for the most part) not taught by professionals: users had to teach themselves, poring over handbooks, manuals, FAQs and documentation, and joining computer clubs.[52]

All this indicates that important shifts are taking place in how people work and consume in the digital economy. But the notion that the Internet is a privileged site of freedom also helps to *justify* the existence of capitalism. Let us first reconsider the ideology's 'cyberlibertarian' roots. John Perry Barlow's *A Declaration of Independence of Cyberspace* declared to governments that 'the global social space we are building [is] naturally independent of the tyrannies you seek to impose on us'.[53] Bodiless cyberspace was the last hope of humanity and freedom. But what was to take the place of governments? The invisible hand of the market, naturally. What Barlow failed (or chose not) to see was that the application of laissez-faire economics and the deregulation of state assets would result in putting the unfettered corporate control of networks beyond discussion.[54]

It therefore comes as no surprise that the Internet ideology is blithely ignorant of the reality of delocalised work in the global high-tech industry, whether at the point of hardware production or at that where the noxious chemicals which constitute it are recycled.[55] The Internet ideology of freedom also champions capitalism in the sphere of consumption. Portraying cyberspace as a cornucopia brimming over with free or pirated content creates in consumers the *need* to purchase the requisite hardware and bandwidth; and it tells them to do so in the name of rebelling against the power of evil corporations, who are intent on protecting their private intellectual property. As the Apple Computer slogan once infamously had it: 'Rip. Mix. Burn. It's your music.' And, of course, the Internet ideology promises to erase the distinction between producers and consumers of content, so that everyone will be an artist or a journalist; a heroically active 'prosumer' or 'produser' instead of an abject consumer.[56]

In this sense the Internet ideology can be understood as a component of a globalised social discourse deployed to motivate capitalism's workers and managers and to ensure its survival. Luc Boltanski and Eve Chiappello have named this process a change in the 'spirit of capitalism'.[57] Capitalism needs to co-opt critique in order to reinvent itself. It is because the 'new spirit of capitalism' has integrated elements of the 'artistic critique' of the New Left, the countercultural desire for autonomy and creativity, that it has been able to justify its amoral purpose – the unlimited accumulation of profit by peaceful means – and motivate managers to embrace it. In the process, it has disqualified the quest for equality and security which proponents of capitalism's 'social critique' (trade unions, for example) had always pursued; autonomy was exchanged for security. The 'new spirit of capitalism' promotes new, liberated and even libertarian ways of making money as well as the realisation of the individual's most personal aspirations. Successful individuals are always busy, always active, drawing no difference between work and play. In a networked universe, where success depends on making connections with interesting others thanks to one's agreeable personality, the 'great man' is autonomous, a nomad, effortlessly shifting from project to project.[58]

The work practices embraced by hackers, such as high productivity, endurance, idiosyncratic reconfigurations of workspaces and unconventional time patterns, are those idealised by the new spirit of capitalism. Similarly, communication and production on the Internet epitomise the informational work ethic: successful bloggers 'post' non-stop – they are always active, always *on*. And creating a personal profile on social networking sites such as Friendster, MySpace, LinkedIn, Spoke or Facebook allows people to demonstrate the extent to which they are popular. On social networking sites the substantive content *is the network* – the nodes, ties and flows – and success is measured by the ever-expanding number of 'friends' (connected nodes). Facebook users playfully showcase to authorised others the links to their networks, as well as the nature of those links, such as virtual gifts, images, causes, games and so on. The public face of Facebook is Mark Zuckerberg, but an equally important figure in the site's development and success is Peter Thiel, a libertarian Silicon Valley venture capitalist who also co-founded the virtual banking system PayPal.[59] But the traditional libertarian concern for privacy has its limits: when it contradicts the profit motive. For the exhaustive profiles listing people's most intimate material, spiritual or consumer preferences – which they have themselves helpfully created – legally belongs to the owners of Facebook, and to the advertisers they sell this information to. In informational capitalism individual users can freely copy and distribute digitised corporate content, and corporations can freely copy and distribute digitised user-generated content.

In this universe, demonstrating that one is both 'sticky' (others respond positively to your requests for connection) and 'spreadable' (others are interested in reproducing your content) acquires the force of an injunction.[60] Those who fail to demonstrate their autonomy, flexibility and resourcefulness, in all arenas of social life but especially in the educational and professional spheres, where competition is particularly intense, those who are perceived as heteronomous, are seen as failures and have no choice save to violently blame their lack of success on others, or to withdraw from the competition.[61]

Countercultural claims for increased creativity and autonomy, far from being aberrations concerning only computer hackers, are the dominant paradigm of today's market economies, which emphasise project and team work, participatory management, computer-supported cooperative work, creativity and reskilling. In the same way, the 'flexibility' that workers will develop through their working careers to accommodate changes in organisational hiring patterns (from full-time lifetime employment toward part-time, contract, outsourced, temporary and casual work) is said to allow them to increase their self-development.

Epistemic Tribal Projects

Having explored the paradoxical genesis of autonomous online activity, it is time to turn to the definition of the phenomenon. Castells suggests that the phrase best characterising people's online behaviour is Barry Wellman's notion of 'networked individualism'.[62] The development of the Internet provides an appropriate material support for the diffusion of networked individualism as the dominant form of sociality, where individuals build their network on the basis of their interests, values, affinities and projects.[63] But can individualism, even when it is networked, really capture what is happening on Wikipedia, for example? The term downplays any notion of collectivism or group activism in online interactions.[64] How then should the social formations created by autonomous agents online be characterised? Historically, an early candidate for such a taxonomy was Howard Rheingold's idea that when enough people carry on public discussions for long enough, with sufficient human feeling, they form webs of personal relationship and become virtual communities.[65] Other authors subsequently expanded on the term.[66] Virtual communities can be defined as follows: communication is their core and definitive activity; membership is voluntary and easily revocable; and the basis of relationships is shared personal interest rather than obligation.[67] However the 'community as communicative process' metaphor is one of convenient togetherness without real responsibility.[68] Indeed, though dialogue and communication are essential, what is occurring online is not just conversation. In terms of the practical purpose of weblogs, wikis and free-software projects, people are gathering online to create something together, to build *projects*.

Engagement in common work points to the importance of the concept of community of practice. A community of practice is a group of people who share an interest in a domain of human endeavour, and engage in a process of collective learning that creates bonds between them, such as a garage band or a group of engineers working on similar problems. A strong motivation for participation is the advancement of competencies by informally sharing practical experience.[69] The concept is especially useful in its elucidation of the way in which new entrants are inducted through processes of 'legitimate peripheral participation'.[70] Learners are encouraged by insiders to legitimately participate in the work of the community; that is, new entrants will not be expected to perform at the same level of competency as incumbents, but be given enough time and support to learn the requisite skills. This also implies that new users should be able to directly observe how core users operate. It is through active and legitimate participation, first on the outskirts of a project, and gradually towards the centre, that education and socialisation occur. A related notion, which has a stronger focus on the cohesion-building internal attributes of informal communities, is that of epistemic community. Epistemic communities are networks of knowledge-based experts who share the same world views, such as principled beliefs, notions of validity, and a common policy enterprise.[71] They provide advice and specific policies for governments and help to frame issues for collective debate. Bloggers and hackers are indeed socialised through a process of increasing participation in a common conversation or project; and wikipedians and hackers are certainly animated by an epistemic concern, and share many of the characteristics outlined above. However, neither concept – community of practice or epistemic community – sufficiently addresses the *affective* quality of Internet sociality. Furthermore, autonomous Internet projects are by definition not oriented towards providing the state with advice, scientific or otherwise.

Which brings us to tribes. The term has flitted in and out of use when referring to modern social formations. Writers in the British cultural studies tradition have put forward the notion of a 'tribal' resurgence. In this line of argument, the deregulation of modern forms of solidarity and identity based on class, occupation, locality and gender has led them to recompose into 'tribal' identities and forms of sociality. Tribal identities serve to illustrate the temporal nature of collective identities in modern consumer society as individuals continually move between different sites of collective expression and 'reconstruct' themselves accordingly.[72] In a similar vein, the French sociologist Michel Maffesoli uses the term 'neo-tribe' to describe new forms of sociality based on *proxemics*, the feeling of belonging. For Maffesoli, the concept of historical centre has exploded into a multiplicity of subterranean centralities which each have their own history and which share an ethos, a way of being together. These tribes may have goals, may have finality; but this is not essential; what is important is the energy expended on constituting the group *as such*. Furthermore, what matters is not so much belonging to a gang, a family or a community but rather the capacity of switching from one group to another. In contrast to the stability induced by classical tribalism, based on ethnically and culturally fixed membership, 'neo-tribalism is characterised by fluidity, occasional gatherings and dispersal'.[73] Within a particular tribe, there are many members who belong to a multitude of other tribes. But why should these associations be called 'tribes'? It could be argued that here 'tribe' is simply a different way of saying 'subculture'. Yet the selection of personas in the capitalist marketplace, a mainstay of the cultural studies vision of subcultures, leads to problematic assumptions. In particular, the celebration of 'resistance' through consumption, when it became detached from the working-class experience in the 1980s, turned into one long celebration of the singular twist that each individual or group could add to the globalised media product – thus legitimising the new transnational consumer ideology.[74]

In short, what is missing from the frameworks sketched above are the specifically *economic* and *political* dimensions of collective autonomous practice. The political potential of the term had been sensed by communication scholars. McLuhan's evocation of a 'global village' was meant to describe a scenario in which electronic media would allow the resumption of earlier, more direct forms of communication. The interdependency of members of traditional tribes, necessary to ensure their survival, was compromised when other forms of communication were introduced.[75] The immediate quality of online communication would resemble the face-to-face dealings of village members, in which, as online, everyone had equal access to all public information. More direct forms of democracy become possible. Communication scholars have explored tribalism in online communities, with studies of far-right identity formation and Nigerian email scamming practices.[76]

These studies show the way, but do not advance far enough. What any discussion of 'online tribalism' requires is a return to the anthropological tradition in order to answer adequately the question: What constitutes a tribe? In the simplest terms, the tribe is the social and political formation which *predates the state*. That tribal organisations represented a kind of 'primordial' or 'primitive' communism was enthusiastically posited by Marx after he became acquainted with Lewis Morgan's accounts of the social and political organisation of Native American groups such as the Iroquois Confederacy.[77] After Marx's death, these ideas were reprised by Engels in *The Origin of the Family, Private Property and the State, in the Light of Researches of L. H. Morgan*. In these accounts a tribal system was seen as a first stage in a necessary evolution towards statehood, as evidenced by the full title of Morgan's book: *Ancient Society or, Researches in the Lines of Human Progress from Savagery through Barbarism to Civilisation*. In the twentieth century, this ethnocentric view of tribes was strongly contested. For the political anthropologist

Pierre Clastres, tribes were explicitly established *against* the state, that is, *against* the existence of a separate organ of power, *against* the distinction between dominants and dominated.[78]

Though affectivity plays a role in the establishment of networks and the recruitment of participants, online tribes are first and foremost social formations which *seek to bypass hierarchical domination*. Tribes favour direct forms of democracy and the pleasurable provision of free gifts, in the context of a shared epistemic project. In economic terms, tribes embrace the production of public goods, as well as the previously mentioned non-monetary 'gift economy' model, with an emphasis on mutuality and cooperation.[79] By making open-hearted contribution to a commons, a mission or a fellowship of which the giver is a part, contributors are reinforcing the conception of the self as part of a collective and of one's effort as part of a collective effort.[80] Since all members are contributors, interpersonal agreement is essential for social cohesion. The attunement to the foundation and identity of the project is essential to guarantee future development.[81] On the Internet, autonomous *projects* become autonomous *tribes* when common purpose and common work lead to autonomous institutions which members use to regulate their work. Beyond the project itself, the purpose is always autonomy. But autonomy cannot account for how participants determine who is telling the truth, or how online autonomists collectively exercise their political will. Tracking autonomous expression and organisation on the Internet requires an investigation of the notion of authority.

2

THE DISTRIBUTION OF CHARISMA

Only hackers can judge hackers.
  Manuel Castells, *The Internet Galaxy*

Authority is justifiable power. The interest for power in being perceived as legitimate is clarified when we think of children: the subordination of children is a temporary state; they are fully expected, having grown up, to be self-reliant and autonomous. This can only mean one thing: subordination is almost by definition seen in a negative light; it requires *legitimation*.[1] The justification of power helps the powerful to gain moral authority and obscure the negative features of power relations, such as exclusion, restriction and compulsion. It is helpful to distinguish between *learned authority* or expertise, which involves buttressing claims to truth, and *administrative authority*, which involves justifiable decisions. The classic account of authority as traditional (hereditary), legal (bureaucratic) and charismatic (revolutionary) was proposed by Max Weber. But is Weber's formulation of authority still relevant in the online context? Authority, and charisma in particular, are useful categories, because they draw our attention to the idea that those who are led (or influenced) *consent* to their subordinate position. But the Internet's socio-technical impact must be taken into account: charismatic authority has been transformed, distributed and aggregated in the online environment.

Sticking it to The Virtual Man

The Internet is widely perceived to be an authority-free zone. To understand the genesis of what is primarily a countercultural argument, it is useful to look back briefly at the group of critical theorists collectively known as the Frankfurt School. For these thinkers, as for Weber, legal authority is not visibly embodied in individuals but is an impersonal bureaucratic force which derives its legitimacy from being universally and consistently applied to all. The Frankfurt School added Freudian psychoanalysis to the critique of instrumental reason and socialisation, as in Horkheimer's *Study of Authority and the Family* which focused on the diminished role of the father in the family under monopoly capitalism. Since impersonal, extra-familial forms of authority now dominate individuals, the internalisation of authority (in the form of a moralistic superego) is replaced by conformity.[2] In *Escape from Freedom*, the psychoanalyst Erich Fromm developed the school's fusion of Freud and Marx, detecting elements of sadism and masochism in social structures such as the classroom, and a desire for adherence to authoritarian hierarchies in the conformity transmitted by mass culture.[3] The school's preoccupation with authority culminated with Adorno et al.'s massive research project *The Authoritarian Personality*, which aimed to assess the tendency of individuals to succumb to intolerance by measuring their authoritarian tendency on an F-Scale (for 'Fascist'). The authors wrote that a 'hierarchical, authoritarian, exploitative parent–child relationship is apt to carry over into a power-oriented, exploitively dependent attitude towards one's sex partner and one's God'.[4] A new anthropological type was unearthed, characterised by rigid conventionalism, submission to authority, opposition to everything subjective, an emphasis on power and toughness, destructiveness and cynicism.

The Authoritarian Personality's methodological and theoretical underpinnings have since then largely been discredited, because of the inconsistent relationship between authoritarianism and childhood experiences, or of the tautology of the 'F-scale' measure of authoritarianism and the attitudes and behaviours it was meant to predict.[5] Nonetheless, the impact was great, and the message – like that of William Whyte's the *Organisation Man* (1956), of Paul Goodman's *Growing Up Absurd* (1960), of Marcuse's *One-Dimensional Man* (1964) – was clear: there was an underlying pathology surrounding questions of authority in both fascism and liberal democracy. Clearly, authority had to be rejected. A report to the conservative Trilateral Commission agreed: the revolts of 1968 represented a general challenge to the existing systems of authority, public and private.[6] The mass artistic critique of authority started with the resistance to the US military–industrial complex and its controversial actions in Vietnam and soon spread to figures such as policemen, judges and fathers; and to institutions such as schools, universities, factories and prisons. During the late 1960s and 1970s, 'question authority' was a popular buzzword, and accusations of being a 'fascist' the worst possible insult. Many believed that social authority was on the wane: Mitscherlich argued that social fragmentation meant that authority becomes vacuous and hence identification to it impossible. In Lasch's view, the implosion of the overarching mechanism of social authority and of the sources of normative identification results in a narcissistic withdrawal towards the self.[7] More recently Toni Negri and Michael Hardt intoned a familiar refrain: disobeying authority, they wrote, 'is one of the most natural and healthy acts'.[8]

How the Internet came to be perceived as a privileged vector for anti-authoritarianism is foretold by a canonical text of new media theory, Enzensberger's *Constituents of a Theory of the Media*.[9] Enzensberger observed that media like television or film do not serve communication, but prevent it. The technical distinction between receivers and transmitters reflects the social division of labour into producers and consumers, or the 'basic contradiction between the ruling class and the ruled class'.[10] Enzensberger drew a distinction between 'repressive' and 'emancipatory' uses of media. Repressive or traditional mass media use can be characterised by centrally controlled programming; one transmitter, many receivers; passive consumer behaviour; depoliticisation; production by specialists; and control by property owners or bureaucracy. In contrast, emancipatory media use is defined by decentralised programming; many-to-many communication; interactivity and feedback; a political learning process; collective production; and social control by self-organisation. Using media in this way will enable the masses to become authors, the authors of history.[11]

This vision has been realised, it seems. Witness Benjamin Barber's opinion that the new media allows ordinary people to bypass the mediation of elites, to challenge hierarchical discourse and encourage direct democracy.[12] Similarly Yochai Benkler describes the move from a 'hub-and-spoke architecture with unidirectional links to the end points' in the mass media model towards a 'distributed architecture with multidirectional connections among all nodes' on the Internet.[13] If everyone has a voice, and everyone can link to everyone else, no one is in a position to dictate what anyone can say or do. Given all this, it comes as no surprise that when it comes to assessing authority on the Internet, a clear consensus emerges amongst commentators of highly different stripes: *there simply isn't any*. Online human groups are stereo-typically presented as less hierarchical and less discriminatory, more inclusive and democratic than traditional communities, where recourse to visual markers of identity often results in prejudicial exclusion, silencing and mistreatment.[14] When a group of marketing and technology luminaries published a manifesto on how to harness the power of the Internet for business development, they declared that the Internet was subverting the unthinking regard for centralised authority, 'whether that "authority" is the neatly homogenised voice of broadcast advertising or the smarmy rhetoric of the corporate annual report'.[15] Founder of the Nettime and Fibreculture mailing lists and prominent Internet theorist Geert Lovink concurred, declaring that 'the Net has questioned authority – any authority'.[16]

Beyond the Internet, it would appear that the network form itself is by definition impermeable to any form of legitimate power. Examples of this heterarchic or anarchic quality range from radical philosophers such as Deleuze and Guattari (and their concept of the decentred and democratic rhizome) to mainstream economists, who assert that networks lack any legitimate organisational authority that could arbitrate and resolve disputes arising during an exchange.[17] For political theorists, networks are especially interesting because – to the degree that they engage truly diverse participants – they must operate according to principles of equality, openness, respect and reciprocity. Standard deliberative virtues are necessary for the network form, with no centralised leadership promulgating goals, norms and strategies to bring participants into line.[18] In the sociology corner, Castells agrees that networking means no centre, thus no central authority.[19] For Boltanski and Chiappello, networks are non-totalisable – that is, not regulated by a general equivalent.[20]

Against this consensus, legal scholar Lawrence Lessig suggests that digital networking displaces rules by codifying them, rendering them more efficient in the process so that, in Lessig's memorable phrase, 'on the Internet, *code is law*'.[21] A law, furthermore, that performatively and quasi-autonomously works to fulfil its function. Following Lessig, Andrew Galloway also argues that the Internet is highly controlled.[22] Galloway draws on Paul Baran's seminal papers on packet switching, which declared that all nodes in the network would be equal in status to all other nodes, each node having its own authority to originate, pass and receive messages.[23] This power, which Galloway calls 'protocol', consists in the scientific rules and standards which allow autonomous agents to operate. Protocol is how control exists after distribution 'achieves hegemony as a formal diagram'.[24]

More prosaically, protocol is what the members of the Internet Engineering Task Force (IETF) agree constitutes useful computer networking technical standards, which are publicised in the form of 'Request for Comments' (RFCs) before being adopted by the Internet technical community. The Internet cannot function without agreed-upon standards which are considered to be authoritative, that is, obeyed by all. This central notion informs a fundamental fact, which can be observed with great regularity: online, *there can be no autonomy without authority*. Far from being anti-authoritarian entities, in a decentralised network, autonomous tribes require authority to perform basic functions: defining what they embrace, and what they reject; what information is relevant or irrelevant; which pronouncement is trusted or distrusted; who is included or excluded. This is, of course, not a very original proposition – in fact, the notion constitutes the basis of most political science courses. But what is new is that, in the course of being networked and distributed, authority itself has been transformed.

Hacking Weber

Max Weber distinguished four types of social action: traditional, affectual, instrumentally rational and value-rational. The first three forms correspond respectively to traditional, charismatic and rational-legal bases of authority, which are meant to convey the myth of the natural superiority of rulers. *Traditional* authority derives its legitimacy from the sanctity of age-old rules and powers, long-established habits and social structures, in the case of hereditary monarchs for example. *Rational-legal* authority depends on impersonal rules and is enforced by professional civil servants. The efficiency of this domination by knowledge renders this type dominant, whether in economic or political institutions. Rational-legal authority is bound to intellectually analysable rules. On the contrary, Weber characterised *charismatic* authority as irrational, and foreign to all rules.[25] Charismatic authority derives from the gift of grace: from a higher power or from inspiration. It rests on the qualities of an individual personality, by virtue of which he or she is deemed extraordinary and treated as endowed with supernatural, superhuman or at least specifically exceptional powers and qualities.[26] At the same time charisma never remains long in its unadulterated form, before being 'routinised' into a more stable form, usually incorporating characteristics of bureaucracy or patrimonialism.[27] Weber was careful to point out that this was no fixed taxonomy, but a methodological device, and that, in practical application, authority systems contain a mixture of these various typological elements.[28]

Weber recognised that 'anti-authoritarian' systems placing a high value on autonomy existed, but he did not systematically analyse the problems of authority and power peculiar to these groups.[29] Joyce Rothschild suggests that collectivist or anti-authoritarian groups are animated by the last type of social action, value-rationality, which has no counterpart in Weber's typology of authority. Value-rational orientation to social action is characterised by a belief in the value for its own sake, independent of its prospects of success.[30] When people's convictions guide their actions, autonomous principles are at work, and there can be no separation between leaders and workers. Collectivist groups such as democratic communes departed from established mode of organisation to such an extent that James Coleman referred to them as *social inventions*.[31]

Authority in online tribes has both an affectual and a value-rational basis: it demonstrates both charismatic and collectivist or *sovereign* characteristics. The countercultural injunction to 'question all authority' really only applied to the rational–bureaucratic variety, with the criticism of the state and corporations (and their reproduction through the education system). The women's and gay liberation movements' challenges to phallo-centrism, and ethnic minorities' rejection of white supremacy were evidently directed against traditional authority. However, 1960s radicals seemed to have no problems with the charismatic aura of assorted revolutionary leaders, gurus and rock stars.

In Weber's classic formulation, charismatic authority is the specifically creative revolutionary force of history, which transforms all values and breaks all traditional and rational norms. Bureaucracy and patriarchalism are antagonistic in many respects, but they both emphasise continuity, whilst charismatic authority and politics emerge in reaction to extraordinary needs, which transcend everyday economic routines. Charisma rejects as undignified all methodical and rational acquisition. Charismatic heroes derive their authority not from an established order and enactments, as if it were an official competence, and not from custom or feudal fealty, as under patrimonialism. Charismatic leaders gain and maintain authority solely by proving their strength in life: prophets must perform miracles, and warlords heroic deeds.[32] Weber's basic premise has been expanded upon by numerous writers: in the 1970s charisma was said to account for revolutionary leadership in social movements.[33] More recently, Shukaitis has contended that Global Justice movement activists use symbols endowed with 'charismatic energy'.[34] The white tunics of the *Tuti Bianche* and the black clothing of the Black Blocs represent a shift in investing a symbolic energy and resonance not in specific individuals, but in the 'social processes and spaces that are created though acts of resistance'.[35] In the field of business management, an extensive literature has analysed the overlap between 'transformational' and 'inspirational' leadership and charismatic authority.[36]

While interesting, these interpretations run the risk of trivialising the term and, in effect, of losing what defines it: the *sacred qualities* of an individual and the *sense of mission and duty* that defines the relationship between individual leaders and their followers.[37] As Teryakian has pointed out, charismatic groups perceive themselves to be moral communities fighting for some kind of transcendence.[38] Refocusing attention on the distinctive nature of charisma does not preclude deploying the notion in novel contexts. Weber segregated the object of charisma, seeing it almost exclusively in its concentrated and intense forms, and disregarding the possibility of its dispersed and attenuated existence. It is more useful to think of charisma as part of a continuum of authority procedures. Since all authority is in a sense 'fallout from charismatic explosions', it can be detected everywhere.[39] Secularised, mediated or institutionalised forms of charisma exist in the modern world, representing the injection of religious or extraordinary qualities into the everyday.

Online charismatic energy is based on extraordinary attributes. Its emergence occurred at the same time as that of what Boltanski and Chiappello call the 'artistic critique' of the inauthentic nature of capitalism and the repressive nature of bureaucratic authority.[40] This coincided with the deployment of 'pseudo-charisma' in the mass media. Pseudo-charisma refers to the wholly rational process whereby bureaucratic or political staffs contrive to imbue (for example) US presidential candidates with manufactured aura through a range of artifices and scripted events.[41] As the 2008 US presidential election cycle amply demonstrated, this analysis has lost none of its relevance: the Democratic nominee, Barack Obama, drew large and enthusiastic crowds to carefully staged events. There the candidate called on voters to aspire to a 'higher, better purpose' in civic life. Other instances of manufactured charisma appear under the guise of the vaunted 'stage presence', 'star quality', 'sex appeal' and 'personal magnetism' of media personalities. Such pseudo-religious experiences are responses to what modernity has brought about, an increasing sense of isolation from others: lost in the 'lonely crowd', people feel the need for connectedness with a leader who unites them with a positive representation of society.[42]

In contrast to pseudo-charisma, the original Internet tribe, computer hackers, constituted a network of autonomous expertise which became the template of all Internet authority: code, and later information, was to be produced, evaluated and disseminated *independently of state or corporate authorities*. Hacker authority is embodied in charismatic individuals who are thought to possess an authentic gift of grace, outstanding coding ability, but it also has several important differences from the Weberian model. First, it does not operate in strict accordance with Weber's definition of charisma as neglecting economic efficiency and rationality.[43] Though hackers were certainly not motivated by the profit motive, they were, as computer engineers, naturally keen to maximise their work to its fullest effect. Theirs was a rational enterprise, albeit not a bureaucratically organised one. In effect, the auton-omisation of expertise resulted in the *separation of rational efficiency from bureaucracy*. A similar blurring of traditional Weberian terminology occurs in the case of meritocracy, which was necessary to attract high-quality contributions from voluntary members.[44] But meritocracy was in a sense charismatised: the gift of grace, brilliance in coding, was rewarded not through a hierarchical process but by the group's affective recognition of outstanding work.

It follows that hacker authority is also different from Weber's characterisation of charisma as not recognising any competing claim, including that of the state, as legitimate. In Weber's view, charisma meant complete surrender on the part of the disciples to the leader. This extraordinary personalistic foundation to charismatic authority constitutes, practically by definition, a threat to the state or any pre-existing legal structures.[45] By contrast, hackers operate as a separate group, operating with its own logic and philosophy, which is nonetheless integrated within a larger ensemble: a tribe within a state. Though it was funded by the government, there was little or no corporate participation in the conception of the Internet's forerunner, ARPANET, which was designed informally and with little fanfare by a self-selected group of experts.[46] Hackers were simply mirroring their scientist forefathers, whose autonomous field (devoted to knowledge production and regulated by rewards of social recognition for contributions of information) was similarly animated by the principle of 'conspicuous contribution'.[47] In a sense, hackers were re-energising the traditional self-understanding of scientists that their activity of seeking crucial truths offers a connection to a transcendent ideal. The esteem of a few and the self-esteem resulting from the publication of scientific articles are only valid because of the belief that something vitally important is being said.[48]

The third, and most important difference from the traditional Weberian typology, stems from the combination of hacker charisma with what Weber called value-rational social action. The affective attachment to a project founder's vision becomes indistinguishable from a conviction-based agreement with the project's autonomous goals; participants then develop an affective attachment to the project. The congregation (*Gemeinde*) that in Weber's view was constituted by a leader and his followers eventually became one with the project itself, and especially with its defence against failure and aggression. This common enterprise, though rational, takes on a transcendent significance: advancing the cause of freedom. When members of online tribes call themselves 'citizen engineers' or 'citizen journalists', they are claiming a connection to a transcendent political value, to the republican ideal. Recognising that online tribes are both charismatic and value-rational is crucial to understanding the seeming contradiction that online authority can be simultaneously concentrated in the hands of charismatic founders and distributed among participants.

The 'founding belief' of the IETF, as expressed by David Clark, upheld the revolutionary dynamic of hacker authority, namely that the legitimate basis for authority is autonomous technical excellence: 'We reject kings, presidents and voting. We believe in rough consensus and running code.'[49] Although IETF membership was in theory open to anyone, the centrality of code running meant that the tribal language was only accessible to a small cadre, excluding managers, politicians ('suits and neckties') and everyone else. Hacker authority is based on deeds, on the verifiable demonstration of skill as well as on a technological version of the gift of grace: outstanding hackers are invested with charismatic authority because of the extraordinary virtuosity they demonstrate when programming. Nothing is as important as the overarching goals of performance and technological excellence.[50] Individuals endowed with exceptional computing skills, such as the MIT hackers of the early 1960s, were portrayed by historians in quasi-mythical terms: 'they were such fascinating people ... beneath their often unimposing exteriors, they were adventurers, visionaries, risk-takers, artists ... and the ones who most clearly saw that the computer was a truly revolutionary tool'.[51]

Though hacker authority is skill-based, skills alone do not suffice: IETF chiefs also derive their authority from personal qualities. Computer hackers were meant to abide by the formal and informal rules of the community, such as refraining from using knowledge or institutional positions for their own exclusive benefit.[52] The allocation of domain names was run by the Internet Assigned Numbers Authority (IANA) under Jon Postel, a computer scientist of 'impeccable integrity', 'the most respected member of the Internet's scientific community', whose management was 'widely recognised as fair, sensible and neutral'.[53] Such individuals have special rights and responsibilities. The crucial operation run by the IETF is the standardisation process. Those institutions and individuals who are in a position to transmute Internet drafts into working standards are endowed with particular prestige. The area of the IETF which oversees the Internet standards process is the Internet Engineering Steering Group, made up of several area directors; the IETF's manual duly tells us that 'many people look on the Area Directors as somewhat godlike creatures'.[54] Chiefs or 'tribal elders' in the free-software community are also described in terms of their extraordinary coding abilities.[55]

Since charisma represents a connection to the metaphysical, it can have no base or earthly motive. On the Internet, the rejection of economic rewards is justified by the fact that it is impossible to compensate everyone in a peer-production model. It therefore becomes much easier to attract contributions if project originators are not earning a profit. In both the IETF and in free-software projects, the non-economic character of the work undertaken is evident. The IETF frowns upon participants to conferences or workshops who display their company's logo, as 'the IETF is about technical content, not company boosterism'.[56] Similarly, in the free-software universe, despite Free Software Foundation founder Richard Stallman's oft-quoted assertion that free software refers to liberty rather than price ('think of free as in free speech, not as in free beer'[57]), honour, respect or 'egoboo' ('ego-boost', in the parlance of science fiction fans), rather than monetary considerations, are what drive developers.[58]

Chiefs Without Authority

When does a project constitute a tribe, and when does a manager become a 'cyberchief'? The answer lies in the presence or absence of salvation narratives, and of strong binary oppositions between good and evil.[59] When autonomous tribal projects are threatened by an evil force which threatens to destroy them, the extraordinary qualities of leaders will allow them to confront this peril and save the day. Zygmunt Bauman argues that the contingency of neo-tribes is a source of danger and anxiety rather than of affirmative potential, as self-defined communities depend on an excluded other, which helps to define group frontiers of belonging.[60] The point had of course long been made by social anthropologists: *there is no easier way of reinforcing boundaries and bonds than having an enemy*. The tribal impulse originates with the construction of a monstrous other. Boundaries are marked because tribes interact in some way or other with entities from which they are, or wish to be, distinguished.[61] Much of what Fredrik Barth observed about ethnic boundaries and boundary maintenance is relevant here. Ethnic identity comes into being and survives through relational processes of inclusion and exclusion.[62] Ethnic boundaries canalise social life: the identification of another person as a fellow member of an ethnic group implies a sharing of criteria for evaluation and judgment; both are 'playing the same game', opening up the potential for their social relationship eventually to encompass many other domains of activity. In contrast, the dichotomisation of others as strangers implies that differences in criteria for judgments of value and performance will be recognised, and that interaction will be restricted.

In the online context, adherence to the principles of hacking encourages computer hackers to create distinctions with external enemies (for example, proprietary software corporations such as Microsoft) and internal enemies (for example, inauthentic hackers or 'script kiddies' who copy others' code rather than creating their own). These monsters' lack of observance of the autonomy imperative renders them inhuman – hence the contrast between IETF committee leaders, who had no enemies to deal with, as computers had not yet been commercialised, and Free Software Foundation initiator Richard Stallman's rejection of polluted proprietary models in favour of sacred and pure free software. Software, information and knowledge must be freely accessible, modifiable and distributable – as argued by Richard Stallman (against Microsoft) and by Jimmy Wales, the founder of Wikipedia (against Encyclopedia Britannica).

Since projects are decentralised, motivation rather than price or commands are used to encourage people to work.[63] This means that online project leaders must endeavour to behave in an anti-authoritarian manner. As far as the personality of leaders goes, the IETF founders were keen to portray themselves as not especially privileged, bringing with them no inheritance of decision-making systems creating endowments or advantages. They described themselves as for the most part graduate students who knew each other, shared professional socialisation and were relatively equal in low status.[64] As Steven Crocker, the author of the first-ever RFC put it: 'We were just graduate students and so had no authority. So we had to find a way to document what we were doing without acting like we were imposing anything on anyone.'[65] Hence the use of the mild term 'request for comment'. The cardinal IETF values are self-realisation and autonomy: though it is safe to question opinions and offer alternatives, 'don't expect an IETFer to follow orders' declares the IETF's manual or *Tao*.[66] In reality, these 'lowly' students issued from elite universities and would rise to the top of the profession. ARPANET was prestigious, and when Tom Truscott and Jim Ellis, two Duke University graduate students, dreamed up in 1980 what was to become Usenet (originally meant as a discussion site for 'UNIX users'), they did it so that non-ARPANET sites could exchange tips and news regarding UNIX.[67] The UNIX operating system exemplified the hacker ethic. Though developed at Bell Laboratories, it was characterised by the rejection of top-down bureaucratic models in favour of collections of tiny, manageable programs under the control of the persons using the system.[68] Usenet embodied much of these values, being – in essence – a modular and evolutionary conferencing system whereby files were copied from one computer to the next.

If autonomy is the source of legitimacy, it follows that online authority must take pains to undermine itself, to avoid appearing heavy-handed and authoritarian. This effect is achieved by the use of auto-ironic or self-deprecating strategies. The best-known examples of anti-authority in the UNIX-inspired free-software tribe take the form of the reinterpretation of a charismatic archetype, that of the messianic guru. Richard Stallman often appears at conferences as 'Saint IGNUcius of the Church of Emacs', wearing a toga and holding a large computer disk platter in the guise of a halo over his head. To join this church, you need only say the 'confession of the faith' three times: 'There is no system but GNU, and Linux is one of its kernels'. Humour is used to spread a serious message concerning property rights and free software.[69]

Another mandatory attribute of authority in the hacker universe is humility: one's work is one's statement, and self-importance is noise which distracts from the only matter of import, code that runs. It is also taboo to attack another's defective work, whilst 'flames' (aggressive messages) about ideological or personal differences are acceptable. Raymond attributes this to the fact that online communication channels are poor at expressing emotional nuance: noise, such as attacks on competence, is more disruptive than elsewhere.[70] Exactly how this differs from academia, where people routinely identify with their work and experience ideological or conceptual differences as personal affronts, is not clear.

The authority of online charismatic leaders derives from an affective attachment based on their extraordinary personal qualities. As we have seen, hacker authority is also based on anti-authoritarian technological merit and on the defence of an autonomous networked project. In addition, online authority is affected by two powerful forces: *distribution* and *aggregation*. In the remainder of this chapter, I examine these forces, and what they have wrought: the emergence of a new form of charismatic authority peculiar to virtual networks, *index authority*.

Distributed Production

Authority on the Internet is not only limited because of a pure concern for freedom: it is limited because leaders have to manage cooperative production. In a free-software project, a number of choices need to be made, such as the selection of members, the allocation of effort (should it be invested in mentoring 'newbies' or in cooperating with other experts?), the point in the process at which decisions about code are taken (early, through the selection of people, or late, through criticism of their contributions?).[71] Lerner and Tirole draw a useful distinction between open-source project leaders' lack of formal authority (they cannot give direct orders to people) and their considerable 'real authority': leaders play a central role in defining initial agendas, revising goals as the project proceeds and resolving conflicts that might result in the 'splintering and outright cessation of the project'.[72] Chiefs in free-software or open-source projects, called maintainers or integrators, need to demonstrate that their project is worthwhile. Success depends on the originators' programming skills as well as on their idiosyncratic personal appeal. Maintainers must prove that they will accept the better patch (contribution), and also give credit where credit is due.[73] Raymond points out that 'it's difficult to convince great people to pay attention to you if you don't have a great reputation'.[74]

The notion of *distribution* plays a key role, indicating that production is decentralised. Information assemblages or compilations have existed in the cultural underground for decades, in the realms of mail art collaboration ('networking') and fanzine production. However, the role of integrators is much more important in vertical or 'cumulatively dependent' assemblages, where elements are interdependent, such as some software systems, than in horizontal collections where elements are complementary, such as encyclopedias or compiled fanzines.[75] The benefits of distribution are numerous. Firstly, provided versions of programs are frequently updated (following the 'release early, release often' model championed by Linus Torvalds), projects improve very rapidly. This is because distributed projects harness the power of peer review, 'autonomous cross-evaluation rather than interventions by external judges'.[76] Instead of two anonymous judges, referees can number in the thousands. The so-called 'Linus's law' states that 'given enough eyeballs, all bugs are shallow'.[77] Online, the process is generally open to anyone who feels confident that their advice will be deemed useful by the tribe.

A second advantage of distribution is that the involvement of multiple contributors with an equal stake in the project's continued existence helps to prevent the hijacking or 'forking' of the project into a different direction than that which it has previously followed.[78] Thirdly, distribution is a powerful incentive for recruitment. The availability of source code makes it possible for an unlimited number of individuals to collaborate in its development. This, in itself, is not enough to guarantee that it will be developed: the possibility must be realised by a community of people willing to invest their time and energy. The capacity of project leaders to successfully attract and retain participants and integrate their contributions is crucial for the survival of the project.

Reasons for participation include demonstrating one's technical competence by measuring one's skill against peers; countering the social isolation of working in top-down organisations on top-down projects; creating useful software; and contributing to a collective enterprise deemed worthwhile.[79] Programmers can also embed in software technical assumptions and decisions that condition the behaviour of future users and developers.[80] In general, participants are not content with simple editing duties (such as fixing bugs, cleaning code, and documentation); soon they also wish to extend the code by adding features, making it more flexible and improving its integration with other components. The distribution of authority is itself the prize sought by participants: a central benefit for contributors to online projects is that *they can themselves assume positions of authority*, for example by taking over a particular module which will subsequently be autonomously developed and articulated to the rest of the project. They have then become co-developers. By definition, those who are not technically proficient have no place in this equation. Since *users* cannot obtain authority their needs will be, for the most part, unrecognised.

Decentralisation reaches its limits when the speed at which a high number of contributions are made requires autocratic resolution. Since these decisions have a degree of arbitrariness the term 'benevolent dictator' is used to describe the integrator's role. The point was made by open-source advocate Eric Raymond during a spirited discussion in a Linux developers' email list over Linus Torvalds's inability to keep up with the flow of patches contributors were sending him (he was ignoring email messages): in terms of the Linux development process, 'Linus is god until *he* says otherwise'.[81] The authority of the project maintainer is the sword that can cut the Gordian knot of endless deliberation.[82] Maintainers need outstanding technical skills to justify their choice of patches and resolve disputes legitimately.[83] In contrast to this unique benevolent dictator model, the Perl community has a rotating cast of benevolent dictators – at any one time one person has the so-called 'patch pumpkin', a mythical token that lets programmers change the source code. In Perl, persons of consequence are 'present or former pumpkin holders'.[84]

The rapidly increasing scale of projects makes it impossible for originators to keep up, compelling them to delegate their authority. Hierarchies of trusted 'lieutenants' with a deep knowledge of the project who share the originator's vision filter patches sent in by untrusted outsiders. This makes it much more difficult for newer entrants to gain authority within projects, and there is less scope for potential challenges to the decisions of integrators. This concentration of capacities in a limited group increases the autocratic power of the leader. Yet authority is based on the ability to convince others to follow; a dictator who alienates his or her followers loses the ability to dictate.[85] In all free-software projects the authority holder's power is held in check by a sword of Damocles, the ever-present threat of schismatic forking, whereby disgruntled dissidents decide to loop off in a different direction, draining off energy and participants. In a sense, forking represents the means to resolve the tension between autocracy and distribution. If the project itself becomes identified as the source of charisma, participants will develop a sense of loyalty to it, and identify with the project more than with the leader, thereby facilitating the process of detaching from the leader. The widespread use of licences such as the General Public Licence naturally aids any contributors who may have 'caught up' with originators to challenge their authority over the direction of the project. Forking creates smaller projects with fewer contributors, and an attendant loss in the efficiency of the project. This draining of resources and weakening of the tribe is usually viewed extremely negatively. For free-software purists such as Stallman, the introduction of open-source licences represents a major fork in the development of free software, a kind of commercial schism. Finally chiefs may cease contributing significantly to the project. Since there is no official contractual basis to positions and relations, sanctions which would subtract authority are difficult to implement. A solution is then to create a new activity which structurally resembles that which is not being carried out, and to let the latter die quietly.[86]

Toadings in the Early Social Net

The Internet is commonly described in terms of sandwiched physical and digital 'layers' enabling people to access the network:

most commentators distinguish a *transmission medium layer* (the cables), a *computer hardware layer* (such as routers), a *software or code layer* (the Internet Transmission Protocols for example) and finally *a cultural or content layer* (actual websites).[87] The content layer within which most people interact comprises a multiplicity of worlds and tribes. Though hacker code and culture still represent a determining influence over Internet practice, the vast majority of Internet users do not program. They are content to use the network's possibilities for autonomous expression and organisation. Vint Cerf, one of the leading IETF chiefs, once noted that most users were not interested in the finer details of Internet protocol: 'they just want the system to work'.[88] The 'Web 2.0' notion that the Internet is 'writeable' for everyone depends on the diffusion of user-friendly tools such as weblogs and wikis, which allow non-technical users to program sophisticated websites and databases.

But well before then, people had been using the Internet for non-programming purposes. Violent flame wars concerning Usenet governance showed that this cohabitation between hackers and non-technical users was not always peaceful.[89] How then did charisma operate in this early social net? Evidently, it still had a strong technical base. Multi-User Dungeons or Dimensions (MUDs) were the earliest kinds of online communities (the first one was formed in 1977), while Bulletin Board Systems and Usenet both began in 1979.[90] MUDs, MOOs (MUDs Object Oriented) or, later, MMORPGs (Massive Multiplayer Online Role-Playing Games) were online universes that exhibited a parodic extreme of charismatic authority. These often comprised behavioural codes enforced by the 'authorities', that is to say by hierarchies of all-powerful 'Gods' and 'Wizards' (administrators). These mighty beings were often treated with exaggerated deference and respect. The Wizard of LambdaMOO, an archetypal MUD, recounted: 'I am frequently called "sir" and players often apologise for "wasting" my time ... I am widely perceived as a kind of mythic figure, a mysterious wizard in his magical tower.'[91]

This deferential treatment was justified in a sense because MUD and MOO administrators could bring down virtual fire from heaven upon users who breached the rules of behaviour. They could 'toad' or modify the appearance of a user's character, 'recycle' or delete the character altogether, and disallow future connections from particular computers which offenders had been connecting from. Expressive displays of power such as public humiliation, punishment and ostracism were of an almost medieval nature.[92] Indices of authority obeyed the same demonstrative logic. Careful attention was paid to the trappings of power, distances were maintained, special spaces created, and signs of distinction carried: one did not enter the private sanctum of a 'God of the MUD' uninvited, for example.

In the 1990s, popular representations of the Internet veered between complete freedom and complete control. Encompassing both ends of the spectrum were the popular Usenet newsgroups, which (by then) could be created by anyone. Depending on whether these groups were moderated or not, they were described as wildly anarchic or tightly controlled. But since this early mass Internet was still directly influenced by the techno-meritocratic ideal, subtle forms of authority were also apparent. On the alt.hackers newsgroup it was necessary to hack the news system to post a message, thereby proving one's credentials.[93]

The Reputation Economy

In online hacker tribes, just as administrative authority is distributed, the strength of individual learned authority or expertise is tempered by its collaborative means of production, in which email lists have played a crucial role. Someone who posts a request for information to a list is appealing to a collective epistemic authority; every member of the list can authorise him- or herself to offer advice without having been specifically asked to do so.[94] People who do not contribute something are labelled 'lurkers', a term with derogatory connotations.

The alt.hackers requirement to hack the system represented an elegant way of resolving a familiar conundrum. The informal and open nature of the network results in the outpouring of a profusion of individual expression. This challenge to the mass media's hegemony over public discourse raises the key question: How accurate are these voices? Speaking authoritatively on a subject must be articulated with another consideration: when everyone can speak, the central point becomes the capacity to be heard – who listens to whom, and how the question is decided.[95] Moving from a mass-mediated to a networked information economy results in a Babel of information overload. Individuals need discrimination; they need quality control. Individuals offering information need to demonstrate relevance; individuals searching for information must be able to determine relevance. When it comes to demonstrating authority, the Internet offers autonomous agents new opportunities to organise and access information. The fail-safe mechanism of wikis documents the process of information creation. More generally, online expression does not necessarily represent a low-quality amateur mimicry of commercial products: it may instead allow more expression and more diverse sources.[96] For example, the traditional mass media cannot provide its readers or viewers access to a 500-page report: the reader is expected to trust the reviewer's integrity to provide a fair summary. By contrast, on the Net, it is possible to link to the report; the ubiquity of storage and communication capacity means that public discourse can rely on 'see for yourself' rather than 'trust me'. The hypertextual nature of the Web allows wikipedians, bloggers and activists to buttress their arguments performatively by directly establishing hyperlinks to relevant information.

The peer production of learned authority also takes the shape of automated aggregation mechanisms. Originally, online status as an expert derived from the learned authority of those who could demonstrate their skill, following the academic model. The introduction of peer production in non-engineering domains such as reviews, news and encyclopedias renders quality more difficult to ascertain. A technical attempt to resolve this tension is the automated aggregation of many individual preferences so 'that authority or truth is now perceived to emerge out of the network, almost as a form of transcendent magic'.[97] The network itself becomes imbued with charismatic potency. Much has been made of the so-called 'folksonomies' produced by bookmarking aggregators such as Reddit, Digg, del.ic.ious, and the like.[98]

The reconfiguration of traditional notions of authority was first illustrated by the user-generated reviewing process popularised by Amazon, which is open to the public. Harnessing the power of community members for quality evaluation of products is not a new idea: members of the back-to-the land movement of the 1970s filled the Whole Earth Catalog with user comments on fertilisers and solar systems.[99] Earlier examples include the science fiction fan community, whose first manifestation was in the form of letters analysing storylines which were gathered at the back of short-story magazines. These letters then split off from the magazines and existed independently as fanzines. Amazon's originality relative to these earlier examples is that it is a mainstream capitalist enterprise that is encouraging these amateur reviewers, who are therefore increasingly challenging traditional cultural arbiters. For reviewers, accreditation is tied to participation: performance within the system matters, rather than external factors such as diplomas. Cases have also been documented of people posting identical reviews for different products in order to earn reputation points.[100] It is the quantity of contributions, not their quality, which matters. A technical fault in the Canadian division of Amazon exposed the identity of these anonymous reviewers. It became clear that not only had a number of authors obtained glowing testimonials from friends, husbands, wives, colleagues or paid professionals: a few had even 'reviewed their own books and slurred the competition'.[101]

Another example of reputation-attribution mechanisms in the commercial world is that of auction sites such as eBay, where the honesty and professionalism of prospective vendors is indicated by the aggregated rankings provided by their previous customers. eBay's reputation system is meant to increase trust in the system as the number of transactions increases. A feature called the feedback forum enables eBay buyers to interact and construct a meaningful history of their seller. There is, however, little incentive for people to complete a survey; because of fear of unpleasantness or retribution it has proved hard to elicit negative comments for moderate levels of discontent; when it comes down to it, there is no means of ensuring honest reporting.[102]

The hacker tribe has built its own meritocratic systems of authority attribution. The most well known is that pioneered in 1998 by Slashdot, a wildly popular free software news hub and the first weblog to allow free commenting. Anyone can contribute stories to Slashdot, but editors decide what appears. Anyone is then free to comment on, and rate, the story. Where Slashdot ups the ante is that the *comments* themselves are rated according to whether the community thinks they are useful. Anonymous comments enter the system rated at zero points; comments signed by new (untrusted) names enter at one point; veterans who have demonstrated their authority in the past enter at two points. Once a comment is made, other previous contributors are randomly selected to assess whether the comment's rating should be raised or lowered. Contributors are allocated five points, with which they can raise or lower ratings by one point only, but cannot moderate posts which they have chosen to comment on. Slashdotters who contribute quality stories and comments on stories are rewarded with 'karma', a time-limited point system allowing them to increase their reputational capital. These moderation points expire after five days, so they cannot be accumulated. Visitors can set user preferences to display only posts with a minimum number of points. The moderation setup is designed to give numerous users a small amount of power, thereby decreasing the effect of malicious or ill-informed users.[103] A similar example of this kind of 'collaborative filtering' is the kuro5hin (pronounced 'corrosion') collective weblog, where all articles go into a moderation queue, and members each get one vote to determine an article's fate. This is the system which the Daily Kos weblog employs to rate posts by anonymous contributors, known as 'diaries'.

This automation of the meritocratic attribution of reputation has led commentators to talk of an emerging 'wisdom of the crowd' or of a 'hive mind'.[104] Yet the mechanical aggregation of independent opinions already lay at the base of the well-documented success of the Google search engine, whose PageRank algorithm aggregates hyperlink patterns.[105] In a data-rich environment such as the Web, the number of pages that could reasonably be returned as relevant is far too large for a human to digest.[106] All sites are equally *retrievable* on the Web, but some are much more *visible* – and hence perceived as authoritative – than others. Being the recipient of links from many sites, especially prestigious or popular ones, is seen as a form of endorsement.[107] For Kleinberg, hyperlinks encode a considerable amount of latent human judgment and this judgment is 'precisely what is needed to formulate a notion of authority' online.[108]

This 'notion of authority' has nothing to do with the legitimate power to rule; but it is more than simply being authoritative, or credible, on a given topic. It represents the crucial capacity to influence, direct or manipulate the terms of the debate by defining the parameters of what is legitimate, worthwhile and interesting. And this is, in fact, another important facet of what has happened to charismatic authority on the Internet: alongside its skill-based variant which occurs in peer-produced projects (hacker authority), online charisma has a component which denotes the central position of nodes within networks, as measured by rankings on search engines such as Google: *index authority*. The term derives from the fact that the authority of an actor is being derived from the relative position in an index of web pages, which is the core component of search engines such as Google.[109]

Not all charismatic agents are human. Bruno Latour has shown that non-human actors are involved in social relations.[110] On the Web, people often treat websites as actors in their own right. Index authority, which emerges from the actions of the network's agents (such as the creation of hyperlinks) can be statistically measured, as in the case of Google's rankings, or of the various sites offering daily readings of the impact of posts made in the blogosphere. Social network analysis has long held that individual centrality in a network is strongly associated with power.[111] In particular, degree centrality (the number of connections received by a node) is seen as a reliable indicator of popularity. The distinction which is frequently drawn in the social network literature between strong and weak ties is less relevant on the Web, where creating hyperlinks between sites represents a trivial cost, and where all hyperlinks are equal. In contrast, it is useful to highlight the difference between degree centrality and betweenness centrality. This latter type of centrality holds that nodes situated between disconnected clusters can play an important connecting role by filling 'structural holes'.[112]

This is especially relevant in the case of high-visibility weblogs, which act as conduits between the blogosphere and the corporate media. That weblogs can play such a connecting role between individual concerns and wider social arenas has comforted the notion, advanced by political theorists and Internet scholars, that the Internet has the potential to rejuvenate the democratic process. For Mark Poster, open online communication would allow Internet communities to 'serve the function of a Habermasian public sphere without intentionally being one'.[113] The original public sphere, emerging in eighteenth-century coffee shops, was defined by Jürgen Habermas as a place where consensus formation rests on the 'authority of the better argument'.[114] What is missing from portrayals of the Internet as a networked public sphere is an understanding of the impact network development has on online communication and organisation. The following chapter addresses structural approaches to Internet sociality.

3

THE TYRANNY OF STRUCTURE

Nothing classifies somebody more than the way he or she classifies.
Pierre Bourdieu, *Distinction*

Contrary to what political theory assumes, networks are not neutral communication spaces. Networks have distinct shapes, properties and impacts. The problem is that these effects generate forms of inequality which run the risk of being totally unrecognised. It is a core principle of social network analysis that powerful actors in networks occupy central positions. But more recently, physical scientists have introduced concepts to explicate the *generation* of centrality on networks. They have suggested that centrality may be a function of the evolution of networks themselves, through mechanisms such as 'preferential attachment'. The first part of this chapter assesses how this explanation for the genesis of index authority impacts the claims of networked public-sphere theorists. But structures run deeper than networks. The second part shows that critical sociology's focus on the socially situated and embodied challenges both network theory and public-sphere apologists.

Power Laws

The notion that some distributions of relationships may be highly skewed, or follow a 'power law' – as opposed, for example, to a bell curve – was introduced by Simon in the 1950s and reprised by physical scientists such as Barabàsi in the late 1990s.[1] Such networks are called 'scale-free' because of their skewed or highly unequal distribution. In scale-free networks, the majority of the sites are smaller than average. Power laws and scale-free distribution are said to operate across a variety of complex systems: from molecules in cellular metabolism to the infrastructure of routers physically connecting the Net. Further, they also apply to social network relationships, such as sexual contact between individuals or shared appearances in movies by Hollywood actors.[2] These networks all have the same structure: an overwhelming majority of peripheral nodes and a tiny minority of hyper-connected central hubs or authorities.[3] Scale-free networks are also endowed with a fractal quality: they look the same at all length scales, whether one zooms in or out.

The World Wide Web in general and the blogosphere in particular, with their wealth of freely collectable nodes and hyperlinks, proved fertile ground for the detection of power law distribution. Researchers noted that there appeared to be a severe imbalance in the allotment of website and weblog linking patterns.[4] Cyberspace was duly characterised as a 'scale-free network' in which some hubs (highly linked nodes) have a seemingly unlimited number of links and are responsible for most connections on the World Wide Web.[5] As for the majority of nodes, they constitute a 'long tail' receiving but a fraction of links and traffic. A study of 433 blogs in The Truth Laid Bear's Blogosphere Ecosystem found a power law distribution in which 3 per cent of the top blogs accounted for 20 per cent of the incoming links.[6] As on the Web, the allocation of inbound links (links *to* a blog) in blogspace pushes users towards small numbers of hyper-successful sites. The power law distribution model has been challenged with the argument that if links patterns are examined within specific communities, such as university or newspaper homepages, they exhibit a more uniform, less skewed distribution model.[7] However such communities all depend on the existence of real-world social networks, which means that members have a high degree of familiarity with one another. Barabàsi points out that, in the online context, this represents an uncommon level of horizontal visibility.[8]

The lack of an explanatory model for the formation and trans-formation of networks is a familiar criticism of social network analysis and its static snapshots of network structures.[9] Barabàsi offers a *generative* model, aiming to explain how index authority is constituted over time. The explanation rests, first, on the growing nature of networks and, second, on the behaviour of new entrants. The number of websites (as of blogs) has been growing exponentially since the start of the Web: as a result, there are many more relatively small young sites than relatively large older ones.[10] Thanks to the growing nature of real networks, older nodes have had greater opportunities to acquire links.[11] But it is the linking decisions of nodes which has the greatest impact on the skewed distribution of links. As is the case on the Web, bloggers know of the most connected sites because they are easier to find. By linking to these hubs in blogspace (sometimes known as the 'A-list'), people exercise and reinforce a bias towards them, a process Barabàsi calls 'preferential attachment', meaning that the rich tend to get richer.[12] In this scenario, linking patterns are an inherently conservative force, leading to the reinforcement of authority. Newer entrants are inclined to link to already well-connected actors, thereby increasing these incumbents' advantage. The disparity of scale between the visible minority and the invisible majority is enormous and growing; there is no way for new entrants ever to catch up. Clay Shirky contends that 'diversity plus freedom of choice creates inequality, and the greater the diversity, the more the inequality'.[13] Indeed, for Barabàsi the most intriguing result of his Web-mapping project was 'the *complete* absence of democracy, fairness, and egalitarian values on the Web'.[14] The Web's topology renders invisible all but a tiny fraction of the millions of existing documents.

The incorporation of *time* as a decisive factor in acquiring authority is not limited to the accumulation of links, but may also be used to describe more intangible notions such as the attribution of esteem or reputation in online tribes. The opinions of those to whom high reputation has been assigned by their peers carry more weight, and as reputation is accumulated over time, the voices of incumbents are the most authoritative.[15] During the previously mentioned email list conflict concerning Linus Torvalds a conciliatory role was played by Ted Ts'o, described by Moody as 'perhaps the most senior Linux lieutenant in length of service, and therefore a figure of considerable authority'.[16]

Free and open-source tribes emphasise the status of charismatic initiators who establish a new and successful project, thereby discovering or opening up new territory. However for most successful projects there is a pattern of declining returns, so that after a while the credit for contributions to a project has become so diffuse that it is hard for significant reputation to accrete to late participants, regardless of the quality of their work.[17]

The Impact of Search

Against the rhapsodies of the Internet's potential for democratic communication, Muhlberger writes that as the number of 'unfit discussants' (people who do not know enough or enter discussion to emote or attack) on the Internet rises, the 'actively engaged' will find Internet discussions less valuable.[18] As the amount of 'faulty information' proliferates, the attention costs of identifying useful and trustworthy information grows, as does the expertise needed to evaluate it.[19] In fact, there exist widespread and convenient means of determining index authority: namely, search engines. However, it has been suggested that search engines may reinforce inequality over the distribution of links and network positions, thereby aggravating the 'rich get richer' phenomenon. Search engines play a critical role in organising access to online information. Without them, the World Wide Web would be like a library containing all the printed books and papers in the world, without covers, and without a catalogue; or a global telephone network without a directory.[20] The power of search engines to highlight certain data and make other data disappear is considerable, as 'to exist is to be indexed by a search engine'.[21] A similar function is accomplished in the blogosphere by ranking sites such as Technorati or The Truth Laid Bear which provide daily statistics of highly-trafficked blogs.

The most successful search engine formula has been Google's PageRank algorithm, initially created by Larry Page and Sergei Brin when they were students at Stanford.[22] Google interprets a link from page A to page B as a vote, by page A, for page B. But PageRank looks at more than the number of votes or links received: it also analyses the page that casts the vote, so that votes cast by pages that are themselves 'important' weigh more heavily and help to make other pages 'important'.[23] 'Important' here stands for: successful at attracting hyperlinks. This has led to the advent of 'googlearchy': the rule of the most heavily linked.[24] The process is similar to the 'Matthew Effect' whereby well-known researchers were found to reap citation benefits more than newcomers.[25] Because the network is so huge, and because no one can possibly cover it all, a cascading network effect determines prominence: the number of links pointing to a site determines site visibility; niches at every level of Internet activity are dominated by a small group of leaders; the dependence on links makes niche dominance self-perpetuating; and because they rely heavily on links, search engines accentuate the rich-get-richer phenomenon.[26] Sites which are linked to by other prominent sites become prominent themselves, whilst others are likely to be ignored; the tendency of surfers to stop after reaching the first returned site with relevant information reinforces this 'winner takes all' phenomenon, resulting in a 'vicious cycle'.[27]

Power laws are not tractable to policy: it is impossible to force people to read what they have chosen not to read.[28] In political theoretical terms, it would seem that the existence of power laws contributes little to the existence or vitality of a 'networked public sphere', a notion which is based, after all, on equality between participants. Yet Yochai Benkler argues that although perfect equality may not exist online, the situation is still democratic, as network-based media offer a new and positively open intake for insight and commentary.[29] It is the attractiveness of the networked public sphere in comparison with the mass-media-dominated public sphere – rather than with a nonexistent ideal public sphere or the utopia of 'everyone a pamphleteer' – that should matter most in our assessment of its democratic promise.[30] Benkler believes that power laws may well have a role to play in online democratic discourse production. This is because high-visibility sites can act as transmission hubs that disseminate information.[31] An example is Sinclair Broadcasting's anti-John Kerry propaganda movie (*Stolen Honor*), which was scheduled to be shown on television a few weeks prior to the 2004 US presidential election. The campaign to protest against this programming, which resulted in Sinclair's cancelling the film's diffusion, was given a strong impetus when a few high-profile weblogs such as Daily Kos sounded the alarm and linked to the unknown weblog which had taken the lead in the campaign. Another central blog, MyDD, provided lists of companies purchasing advertising space on Sinclair's networks, as well as tips on how to frame complainants' views when calling these companies.

In short, the network's clustered and interconnected topology allows the rapid emergence of a theme, its filtering, synthesis and rise to salience.[32] An analysis of the blogosphere came to a similar conclusion: information tends to filter towards the top.[33] The fragmentation of private agendas that never coalesce into a platform for political discussion is thereby resolved.[34] Index authority stabilises and organises democratic discourse, providing a better answer to the fears of information overload than either the mass media or any efforts to regulate attention to matters of public concern.[35]

This seems politically efficient, but does it answer the charge that scale-free distributions are undemocratic? Shirky zeroes in on the issue of justice when he asserts that, given the ubiquity of power law distributions, asking whether there is inequality in the blogosphere, or indeed in 'almost any social system', is the wrong question, 'since the answer will always be yes'.[36] Shirky believes that a better question to ask is: Is the inequality fair? And he believes it is, for four reasons: first, weblogs cost nothing to produce, and there is no vetting process, so it is only marginally harder to have a weblog than to be online at all; second, blogs must be updated every day or risk 'disappearing'; third, stars exist not because of cliquish behaviour, but because of the preference of hundreds of others pointing to them; fourth, there is no A-list, because, other than the distinction between the first and second position, the line separating top nodes from others is arbitrarily decided: there is no qualitative difference between the A-list and their immediate neighbours.[37]

Unfortunately this reasoning exhibits the ignorance of history that characterises many interpretative schemes which hope to distil social relations into neat formulas, be they based on mathematical equations (as with economics or physical science) or on pure disembodied logic (as with political philosophy). It adopts the *structuralist instrumentalist* bias of some network analysts, who believe that actors are utility maximisers pursuing their interests in money, status and power in 'precisely the ways predicted by theorists of rational choice'.[38] Such accounts ignore the impact that history has on relations of power. Shirky concedes that an amount of stasis may emerge, making the structure resistant to change.[39] But he does not consider the possibility that not all nodes on the network start out with the same resources, or are evaluated in the same way.

The Persistence of Archaic Force

Just like public spheres, networks are not neutral communication spaces or level playing fields. Preferential attachment does not explain why certain actors experience more success than others in fostering connections and accumulating resources. Like some strands of social network analysis, it portrays the activity of social actors in narrowly instrumental terms and neglects altogether the impact of actors' beliefs, values and normative commitments.[40] In addition, preferential attachment fails to account for a number of important questions: the inheritance of advantage (the 'silver spoon' effect), the transference of advantage from one field to the next, and the role of classification (who is allowed to participate, and which issues are seen as significant and important). Cultural factors constrain and enable historical actors, in the same way that network structures themselves do.

Members of the Internet Engineering Task Force comprised a self-selected oligarchy of electrical engineers and computer specialists constituting a homogeneous social class: highly educated, altruistic, liberal-minded science professionals from modernised societies around the globe.[41] A statistical study by Lakhani and Wolf, which surveyed developers from a random sample of open-source projects on Sourceforge.net, found that a solid majority of contributors were experienced, skilled individuals with jobs in the technology industry. The average contributor had more than a decade of programming experience; 55 per cent worked on open-source projects as part of their job.[42] Surveys of the blogosphere reveal a similar pattern. An analysis of top A-list bloggers reveals that they are not only white, male and middle-class: they are also highly educated, placing them effectively higher on the social ladder than the 'elite' mainstream journalists whose power they are supposed to be contesting. More than half of all blog traffic went to bloggers with a doctorate (JD, MD or Ph.D.). No other part of media is so skewed towards the elite, towards people who write for a living.[43] As in the Athenian agora, many voices are left out of this 'democratic debate'.[44]

A fundamental principle, the heart of social domination, is at work here: *making the socially constructed appear natural*. Every established order tends to produce the naturalisation of its own arbitrariness.[45] Why are some people so incredibly successful at fostering connections, or enjoy privileges that they have not necessarily earned? The reason is that such people are better equipped than others to accumulate connections because they have inherited resources – whether economic, cultural or social – which give them an advantage over others. Further, these inherited forms of capital or power explain why the products and identities of some people are considered valuable: because they are most in tune with the dominant values of the social space.[46]

What this means is that beneath the network (and the public sphere) forms of non-democratic power persist. In other words, index-charismatic authority rests on *archaic* foundations, because the advantageous quality of being an 'early entrant' is system-atically and unfairly allocated according to age-old inequalities in gender, class and ethnicity. Feminist critics of the idealised public sphere such as Nancy Fraser have noted that we can no longer assume that the bourgeois conception of the public sphere was simply an unrealised utopian ideal; it was also a masculinist ideological notion that functioned to legitimise an emergent form of class rule.[47] New gender norms advocating female domesticity and a sharp separation of private and public spheres were key signifiers of bourgeois differences from both higher and lower social strata.[48] Though Nancy Fraser explicitly refers to the ideas of the French critical sociologist Pierre Bourdieu, such references are few and far between amongst feminist and radical political theorists. This is not surprising, as Bourdieu's point of view was explicitly geared against abstract philosophy and bodiless networks. He wished instead to expose the social and economic inequalities, the strategies of cultural restriction and exclusion, which undergird the everyday.

Field Logic: Bourdieu

Bourdieu rejected the abstract formalism of theoretical theory and embraced a critical approach. Critical theory can be defined as a project of social theory that simultaneously undertakes a critique of received categories and a substantive analysis of social life in terms of the possible, not just the actual.[49] Critical sociology thus aims to uncover the effects of the domination imposed by a number of institutions (such as corporations, and the education, justice and police systems) in the service of a state whose mission is to entrench and reproduce this domination in favour of the social groups which control this state.[50]

In reaction to the strictly Marxist conception of ideology as false consciousness hiding the reality of social relations, Bourdieu insisted on the structuring capacity of culture. Within certain limits, symbolic structures have a constituting power; culture can literally shape reality because the associations and dissociations that it actively formulates contribute to creating social structures – in particular, by legitimising or misrepresenting the political power which creates social structures of domination.[51] Hence Bourdieu's concern with *distinction*, or legitimate taste. For Bourdieu, sociology was never more like social psychoanalysis than when it confronted taste. This was a vital stake in the struggles fought in the field of the dominant class and the field of cultural production – here, sociologists found themselves in the area par excellence of the denial of the social.[52] Distinctive taste serves to reproduce social hierarchies because people are conditioned to embrace the values, cultures and social trajectories which are assigned to them as appropriate.

Bourdieu's theoretical model was built against what he viewed as the excesses of structuralist theory; though his is by no means a 'philosophy of the subject', his focus was on the economising strategies of agents. To this end, the notion of 'capital' (or power) was expanded to include both material and non-material phenomena. Bourdieu focused on the relationships of power that constitute and shape social fields. In anthropology, fields have been defined as aggregations of relationships between actors competing for similar prizes or values.[53] In Bourdieu's sociology, society is differentiated into a number of semi-autonomous fields, internally coherent microcosms, governed by their own 'game rules', yet with similar basic oppositions and general structures.[54] Fields are distinguished according to the kinds of specific capital that are valued in them: capital is 'heteronomous' (external to the field), or 'autonomous' (unique to that field). Fields also differ according to their degrees of relative autonomy from each other and in particular from the dominant political and economic fields.

Bourdieu called charismatic authority *symbolic capital*. The power to impose on other minds a vision, old or new, of social divisions depends on the social authority acquired in previous struggles: symbolic capital is a credit, the power granted to those who have obtained sufficient recognition to be in a position to impose recognition.[55] Symbolic power is the power to make things with words, to constitute groups, to become a spokesperson after a long process of institutionalisation. When Markos Moulitsas of the Daily Kos weblog conferred a hyperlink, and a supportive commentary, on the website which ran the campaign to challenge Diebold electronic voting machines, he was using his power of consecration or revelation to distinguish a group from other groups, or reveal things that are already there, but existed hitherto unseen.

If Bourdieu had spent any time thinking about the Internet (which he didn't), he might have said something like this: What we have here is a classic example of a para-artistic autonomous field with its specific forms of capital and anti-economic rewards. Incumbents are highly educated white males who dominate others thanks to this so-called common good, 'free software', which operates as a covert defence of their own interests as the exclusive repositories of technological expertise. Coding for code's sake also allows these persons to profit from the interest in being perceived as disinterested. Hackers monopolise the power to say with authority which persons are authorised to call themselves hackers, hence the disparagement of 'script kiddies' or 'hacktivists', who are not interested in programming per se, but in the use of applications for fun or activism.

Online authority, Bourdieu might have continued, should be understood as an archetypal example of the values of the dominated fraction of the dominant social group – that is, people endowed with intellectual rather than economic power. This explains the ambiguous view of economic success in hackerdom, and by extension in all online tribal projects. These projects are structured by a rejection of 'corporate' values, yet reproduce the advantage of those who are endowed with various forms of capital, particularly cultural capital. There is more than a little truth in such an analysis. The non-programming early Internet was strongly imbued with a logic of distinction: possessors of exclusive email addresses such as the WELL (Whole Earth 'Lectronic Link, one of the first alternative online communities), or research universities, were viewed in a better light on Usenet than users with commercial accounts. The profusion of terms developed in computer-mediated communication environments to describe the implementation and violation of behavioural norms ('netiquette') points to the importance of cultural capital on the Net. The boundaries of the hacker field are much easier to protect than those of the blogosphere, as blogging does not require esoteric technical knowledge.

The Internet is – almost by definition – an exclusive enclave. People preoccupied with economic survival have more urgent concerns than going online. A recent British study of how people use the Internet to find political information or contact politicians showed that, whilst highly educated people were more likely to engage in offline participation than less educated individuals, they were even more prominently represented when these activities moved online.[56] Bourdieu posited the existence of a mechanical relationship between hierarchies of taste and oppositions between social fields (such as the dominant and dominated fields) or between dominant and dominated fractions of the dominant field. He asserted that these hierarchies reproduce social inequality between a cultured dominant class, a lower middle class with cultural aspirations, and a dominated class which is kept apart from high culture.

A new generation of sociologists has questioned some of Bourdieu's assumptions. For example, Bourdieu's rigid segmentation of high and low (or legitimate and non-legitimate) culture has been criticised by Bernard Lahire. Lahire shows that the distinctive practices of the great majority of individuals are 'dissonant', mixing highly legitimate and non-legitimate activities, rather than 'consonant', that is to say constituted of homogeneous cultural practices. According to Lahire, 'consonant' practices are only found at the extremities of the social field, in the highest and lowest social groupings. The more common situations, those of agents whose relationship to cultural distinction is divided by inner contradictions, are located in the middle echelons of society.[57] These are exemplified by the participation in online tribes, with their mix of the highly specialised (the high literacy of bloggers, technical expertise of hackers, encyclopedic knowledge of Wikipedians) and the intensely quotidian (the lack of formality in exchanges and frequent use of profane language).

Gendering the Online Abject

However, mixing codes is not open to all, and in fact it is precisely through the entitlement to the cultural properties of others that new class relations and new forms of exploitation are entwined and coproduced. The British sociologist Bev Skeggs has shown that what are resources for one serve to essentialise and pathologise another.[58] The propensity for cultural mobility is the preserve of the middle classes, who can playfully dip into and mix class attachments, unlike the working class, whose attempt at 'doing middle' results in failure or pretentiousness; so that it is not simply a matter of the powerful claiming marginality, but of the powerful displaying their mastery of power by playing with it: 'a matter of having your authority and eating it'.[59] This serves to enhance the value of personhood. For Skeggs, the cultural resources for self-making and the techniques for self-production are class processes and making the self makes class.[60] The problem with dominant bourgeois models of the self is that they present the working class as an individualised moral gap, as a failure of the self to know, play, do, think and/or repeat itself in the proper way.[61] Today it is the white working class that is 'abject', the visceral site of psychic and political, hence physical and metaphoric, disgust; of the nonhuman; marking the limit of proper personhood, obese and beyond recuperation.[62]

It is important to recognise that abjection has a precise function in online tribes, which is peculiar to the Internet environment. Though as a polluting influence it is banished to the fringes, it should be differentiated from practices such as trolling (making provocative statements) on Debian lists, or using sock-puppets (fake identities) on Wikipedia, for example. These practices incur banishment, but they are not abject: such vandals are not excluded because of their *innate characteristics as persons*, but because their *behaviour* violates the norms that guide online behaviour. In order to define online abjection, it is useful to remember what online epistemic tribes are: self-organised communities of experts. It logically follows that different varieties of *cluelessness* (lack of expertise) define the online abject.

The most easily identifiable abject individual is the hapless *noob*. In the 1990s, members of the IETF were worried about the influx of newbies – those from AOL appearing to be 'especially clueless' – who would need to be 'socialised' or educated in the arcane lore of the Net.[63] Newbies will eventually become integrated, or drift away. A more serious kind of transgressor will always be tarred with the brush of abjection: those who break the rules of ascetic disinterest in order to accrue personal gain. By doing so, they shatter the illusion of online autonomy's distance from base motives. To gain further insight, we need only examine the terms used to characterise those who engage in such behaviour: on Slashdot, contributors who post vapid comments solely designed to obtain 'karma' points, are referred to as *karma whores*, whilst in the blogosphere the practice of trawling for hyperlinks from prestigious weblogs is commonly referred to as *linkslutting*. Is it a coincidence that practices contradicting the spirit of online autonomy are described with terms describing female prostitutes, the *embodied abject*? It is not. Cyberculture scholars posited that the anonymity of online communication made it gender-blind, or that virtual embodiments allowed all kinds of novel combinations of sexuality and gender.[64] But other early analysts of online sociality had observed that much more than class, *gender* is the one characteristic of our embodied lives that is a central feature of online interaction.[65] And though a decade has passed since then, the sad fact is that sexism is alive and well on the Internet. This is not so surprising: the archetypal archaic power, from which, perhaps, all others flow, is male domination.

Masculinity is an eminently relational notion, constructed in front of and for other men and against femininity, as a kind of fear of the female, and firstly in oneself.[66] Gender is a principle that lies behind the series of oppositions that structure society, even though the specifically gendered nature of these oppositions may not always be fully recognised. The fundamental opposition between masculine and feminine is instead 'geared down' or diffracted in a series of oppositions which reproduce it, but in dispersed and often almost unrecognisable forms.[67]

The hacker focus on technical skills reflects the gendered vision of human activity in which the male life of the mind is valued over women's confinement to the visceral body, and which excludes females from a technological sphere peopled by male 'nerds'.[68] Males have more time and less guilt when devoting themselves to the pleasurable, frivolous, and non-productive aspects of technology, such as gaming.[69] Perceiving technology as a toy serves to position oneself as a man, while perceiving technology as a tool serves to position oneself as a woman.[70] This assertion of gender identity effectively excludes women from the most intimate and distinctive relations to technology. They will not access the economic, social and cultural advantages deriving from high-level computer use.[71] On the Internet, gender bias is apparent when one considers how much more attention free labour attracted when it involved open-source software creation than mailing list and website maintenance.[72] The prestige afforded to early entrants is another way in which male privilege is naturalised and hidden: when the road rules for the superhighway, such as netiquette, were being worked out, there was no critical mass of women involved to ensure that the highway code reflected some of their priorities and interests.[73]

That online discourse and communication is gendered has been a constant finding of Internet research since the 1990s. Susan Herring, the principal voice in the field, found that in many respects the Internet reproduces the larger societal gender status quo: top-level control of Internet resources, infrastructure and content is dominated by men; and one of the main sources of revenue on the Internet, the distribution of pornography, is not only controlled by men but depicts women as sexual objects to be consumed.[74] In terms of online communication, Herring defined early on a series of oppositions characterising the female/male divide: attenuated assertions versus strong assertions; apologies versus self-promotion; explicit justifications versus presuppositions; questions versus rhetorical questions; personal orientation versus authoritative orientation; supporting others versus challenging others.[75] These distinct modes of communication have contrasting purposes: by aligning themselves with and expressing support for others, women create solidarity and promote harmonious online interaction; whereas by challenging and criticising others, men attract attention to themselves and engage in 'contests', as a result of which they lose or gain status.[76] In general women have a deep aversion towards the kinds of adversarial exchange that men thrive on. The constancy of this dichotomy also helps to explain how anonymity online is easily uncovered. An analysis of discourse on academic email discussion lists showed that women had to combine elements of 'men's language' (so as to be taken seriously as academics) with elements of 'women's language' (so as to avoid being considered unpleasant or aggressive).[77]

Other researchers have also determined that males on lists are more likely to assume an initiating role, rather than a responding one; when they do respond, they are less likely to address their respondents directly, preferring instead to use their response to address the audience in general, claiming the topic for themselves. In other words, they are more likely to present themselves as authorities on a given topic. Females tend to address others more directly or call on their attention. This may indicate a desire to signal alignment with others on a list or to claim familiarity with them, as a way of gaining membership by acquiescing to conventions set by those with high status.[78] Gender stereotyping has a clear effect on communication. The expectation that women should talk less, and in a less confrontational manner, amounts to a highly effective censoring of their views. Conversely, the logic of defending one's honour at all costs, of not backing down when challenged, lies at the root of the masculine ethos. Online, this atavistic propensity is couched in the noble civil libertarian tones of early Net adopters. Autonomy from censorship and agonistic debate as the means to advancing knowledge are at the core of this belief system. The male adversarial style is reflected in the early sets of norms of Internet behaviour: netiquette only discouraged 'flaming' (aggressive emails) if it was for personal reasons.[79] Attacking or deriding someone's ideas or values was tacitly permitted, as it conformed to the ideal of proving one's valour in combat, or served a disciplining function. 'Trolling' removed this function, so that only malicious confrontation remains. Usually beginning in an uncontroversial manner, trolling entails luring others into pointless and time-consuming discussions.[80] The preferred aim of trolling is catching inexperienced users. New users tended to be women, the young and other non-traditional computer users: significantly, a Troll FAQ author referred to the generic target of trolling as 'she'.[81] The notion that hostile or harassing speech should be protected by the US Constitution's First Amendment, which guarantees freedom of speech, was originally a contested notion; in fact, women who rejected it were often accused of 'censorship'.[82] However this idea has since acquired near-hegemonic status and is being voiced by both women and men.[83]

Online, the clearest expression of symbolic violence is the definition of the respected subject and its antinomic abject. Discussions of what is of value in the blogosphere – and more generally online – invariably conjure the ultimate abject, the *teenage girl*, a figure which seems to crystallise all that the hacker abhors: technical incompetence, a frivolous lack of concern with weighty matters of protocol, policy, and politics, in favour of the intimate, the trivial, the gossipy, the revelatory. In the free-software universe, this abject does not exist. In the blogosphere, it is excluded beyond the boundaries of the genre: female personal weblogs are not 'weblogs' at all, but 'journals' or 'diaries' dealing with interior issues. 'Weblogs' deal with politics and technology, with the agora.

Despite his well-publicised hostility to postmodern theory, Bourdieu's interest in dispelling the opacity of historical processes is remarkably similar to that of the French philosophers who premise personhood on the experience of forces beyond the control of the subject.[84] Mechanisms such as technology as pure spin (Virilio), technology as simulation (Baudrillard), technology as desiring machine (Deleuze and Guattari) technology as state-scientific control, or as subjectivity (Foucault): they all offer faint echoes of fields as sites where people must engage in strategic accumulation. The aim, always, is to deconstruct the way things are done through people, and, particularly, through their bodies. But can it really be that people, when confronted with the myriad conflicts of life, never distinguish right from wrong, and act on that intuition? The challenge for social research is to take into account both the need for a sense of what is just, as well as the persistence of archaic forms of power behind online charismatic authority.

4

THE GRAMMAR OF JUSTICE

Because it can express itself only in general and abstract laws, the united will of the citizens must perforce exclude all nongeneralisable interests and admit only those regulations that guarantee equal liberties to all.
Jürgen Habermas, *The Structural Transformation of the Public Sphere:*
An Inquiry into a Category of Bourgeois Society

While its usefulness and creativity is undeniable, Bourdieu's approach also generates questions. What is missing from critical social science's vision of ordinary people? Proponents of 'pragmatic sociology' argue that critical sociology neglects an important driver of social interactions: people's conscious use of justice-seeking mechanisms. This chapter suggests that critical and pragmatic approaches are not necessarily antagonistic and could in fact fruitfully enrich one another. This point is illustrated by the relationship of the two axes which structure the space of online tribes: charismatic and sovereign authority. In order to understand the latter concept, this chapter also focuses on the development of self-determined rules, norms and governance procedures on the Internet.

Criticality and Justification

Bourdieu set out to discover general laws of fields or transhistorical invariants; that is, sets of relations between structures that persist within circumscribed but relatively long historical periods. This means that little latitude is given to the internal transformations of social systems. But Bourdieu assumed, rather than empirically demonstrated, a high level of resemblance between fields.[1] Bourdieu's vision of the social world is an economics-inspired approach in which people use utility-maximising strategies to accumulate resources. This world is constantly animated by the competition for power. Can there exist a field not structured by hierarchy, male domination, cut-throat competition, symbolic violence? Apparently the game never ceases: non-competitive solidaristic values do not compute. The inexorable quality of the hegemony of power leads to an immobilisation of history. A way out of this impasse is to change perspectives, both in terms of the analytical categories used and of the researcher's point of view on 'ordinary people'. In terms of analytical categories, the stated purpose of critical sociology, unmasking the machinations of the 'social unconscious', necessarily implies the existence of an objective evaluation standard. In other words, unveiling social domination resulting in unequal distributions of material and immaterial goods implies that one can conceive an alternative model of distribution of these goods. For Boltanski and Thévenot, political involvement must be based on principles of *justice*, employed to evaluate whether a social situation is acceptable.[2]

In addition, the psychoanalysis-inspired notion of cultural unconscious does not mesh with people's observed capacity for self-reflexion. Critical sociology does not sufficiently account for the critical operations undertaken by actors. People are not cultural dopes who lack insight into the normative underpinnings of their actions.[3] People are endowed with *reflexive and critical capacities* (which are not necessarily expressed in public) which question the exteriority of sociologists as sole possessors of truth. They use arguments which display similar features to sociological or scientific reports: valid arguments rest on a system of proofs, on the selection of pertinent facts, on unveiling operations. It is impossible to maintain a radical distance between the everyday activity of 'ordinary people' and the scientific activity of sociologists.[4] For all his talk of the need for social scientists to exercise reflexivity, this is not a field in which Bourdieu particularly distinguished himself. He was not alone in this respect, of course. One is reminded of what De Certeau had to say about Foucault's notion of the unconscious structuring force of the *episteme*: 'Who is he to know what no one else knows, what so many thinkers have "forgotten" or have yet to realize about their own thoughts?'[5]

Bourdieu did argue that socialised subjectivity or *habitus* is a structure both structured and structuring: people's dispositions are conditioned by their field positions, but they can (up to a point) exercise self-determination. Yet despite his aversion to empty 'grand theorising' and professed sympathy for the attention to local detail of 'micro-sociologies' such as ethnomethodology or symbolic interactionism, Bourdieu identified too deeply with the critical viewpoint to really question its central assumption: that sociology alone has the theoretical capacity to unmask 'relations of force that are not immediately perceivable'.[6]

Luc Boltanksi's 'pragmatic sociology' was inspired by the symbolic interactionism of George Herbert Mead and Herbert Blumer, which holds that human groups and society fundamentally exist in terms of action.[7] The logic of action should be understood in the broadest sense possible: constructing a theory, justifying oneself, associating with others, and failing to act, are actions. The critical stance is defined by the action of *unveiling* hidden reality, and of *denouncing* injustice.[8] The pragmatic approach sees the world as overflowing with a multitude of beings, sometimes humans, sometimes entities, which never appear unless the state in which they occur is simultaneously described: there are no persons outside actions.

Boltanski and Thévenot attempted to map a comprehensive set of the forms of common goods which are usually referred to in our society. They contended that people seeking agreement focus on a *convention of equivalence* external to themselves.[9] They defined 'cities' to model these equivalence principles, these ways of ordering people, and examined *appeals to justice* in conflicts where people exercise critique or seek legitimate solutions. This approach was inspired by Michael Walzer's idea that there exists a plurality of regimes of justice.[10] It is of course possible to argue that the value system which underpins the 'civic city' (the common humanity of all people) 'trumps' all the others. Should not all the equivalence principles be measured on a common standard, that of equality?[11] Even though some justificatory regimes may ultimately dominate others, pragmatic sociology's contribution is to show *that there are several such regimes*.[12] In other words, since different orders of justification can always be invoked, there can be no unique normative connotation of public space as enhancing rationality.[13] From the perspective of analysing domination in autonomous online tribes, the focus on the action of justification also offers a useful corrective to critical sociology's insistence that domination is always the reproduction of illegitimate advantage: people can refer to commonly understood ways of ordering the legitimacy of individuals and websites.

As defined previously, the online charismatic justificatory regime is based on anti-authoritarian meritocracy and autonomous genius on the one hand; and on the aggregation of decisions by individuals which result in a node acquiring a central or bridging position on a network on the other. What these divergent manifestations have in common is that they are determined by individual choices – and underpinned by masculine domination. That archaic forms of power persist in new social fields such as the Internet lends credence to the existence of *anthropological constants* structuring social interaction. When seeking to determine what other modes of authority people mobilise in online contexts we can therefore look to the authors who have detected two main modes of human structuration, one centred around individual autonomy, the other around more collective and community-oriented concerns.[14] Interestingly, these constants match precisely David Beetham's contention that two legitimising principles are more emancipatory than others: the principle of individual merit, and the principle of democratic sovereignty, based on the collective will of the group.[15] These two axes divide the field of online tribal authority: the axis of charisma is intersected by an axis of *sovereign authority*.

Legal Autonomy and Sovereign Authority

When people can justify their actions by referring to a formal social contract, we are in the presence of an authority which derives its legitimacy from the general will. Unlike Montesquieu and Locke, for whom democracy depended on the existence of representative institutions, Rousseau emphasised direct democracy and popular autonomy, the equal participation of each person in the practice of self-legislation.[16] Sovereign authority exists when the notion that participants, thanks to their contributions, have a common ownership of the project is given a *material* basis, through institutional forms such as electable arbitration committees which can make or interpret predictable and enforceable *rules* about the propriety or impropriety of certain actions. Rules, which are prime mechanisms for the deployment of authority, serve two important purposes in online tribes. First, they help to define who is part of the tribe. The territory of their extension constitutes the boundaries of the tribe, and exclusion from the tribe is the ultimate disciplining measure. Secondly, and most importantly, rules ensure that the routinisation of charismatic authority is *democratic*.

Usenet was the site of one of the first conflicts between ordinary users and the 'sysadmins', who controlled the various university and private computer systems which hosted the newsgroups. The sysadmins who were most influential in shaping the network's evolution were known as 'net.gods'. Seeing a rising tide of non-technical newsgroups whose users were demanding a say in the running of the network, a sysadmin posted a rant which began with 'Usenet has no central authority... in fact it has no central anything'.[17] He went on to declare that Usenet could not be democratic since it was not an organisation, and only organisations can be run as democracies, as disorganisation means that enforcement mechanisms are impossible. Albert Langer replied for the users: 'I always get suspicious when somebody says "there are no authorities here". My suspicion is that there is indeed an authority but that it does not welcome scrutiny.'[18] The autocratic power of sysadmins to accept or reject the creation of whichever newsgroup they deemed acceptable was eventually blindsided by the emergence of the GNU/Linux operating system, which enabled the democratisation of the technical means to create Usenet groups.[19] The conflict exemplifies a central tension in online tribes between the bottom-up desire for self-determination of ordinary users and the hard fact that hacker expertise in effect determines administrative authority, the capacity to control servers and hence people's online presence and actions.[20]

For Raz, a legal system performs three basic activities of governance: rule-making (legislation), rule application (adjudication) and enforcement (policing).[21] Rules are a practical necessity. But as was observed about MOOs in the 1990s, law also serves a symbolic function, signifying that online projects are more than games. It is not just about recreation, but also about the creation of a virtual polity: 'games have rules, but who ever heard of a game with a Supreme Court and a complex legislative system?'[22] Much the same could be said about the legitimising role of law in contemporary online tribes.

To understand how rules operate in online tribes, we need to recognise that there are different conceptions of the law. In the traditional Austinian conception (named after the English legal philosopher John Austin), law is imposed upon society by a sovereign will which acknowledges no superiors. It is backed by threats, and directed to a population which provides obeisance. A more sophisticated form of the conception of legal power as a tyrannical force draws on the work of Michel Foucault. Surveillance and disciplinary practices are said to make repressive power immanent in society. Discipline involves multiple methods of regulation of individual behaviour, from workplace time-and-motion efficiency directives to psychiatric evaluation; and the classic theory of sovereignty and the legal code centred around it serve to conceal discipline. In this Foucauldian perspective, regulation on the Internet operates not through law but through *practice*, inasmuch as the 'state has worked actively to embed or hardwire the legal regime in the technology itself'.[23]

In essence, this has become the dominant view of Internet law, as formulated by Lawrence Lessig: people are oriented online by a combination of law, markets, social norms and architecture (or code), but it is the latter element which dominates. Code as law means that 'effective regulatory power [shifts] from law to code, from sovereigns to software'.[24] At the same time, commerce requires the identification of identities. The risk is that this will diminish autonomy: left to itself, cyberspace will become a perfect tool of domination, especially in the realm of copyright control.[25]

Before the 'code as law' paradigm became orthodoxic, there was in fact great interest in the legal community in the autonomous emergence of law on the Internet. Legal scholars could draw on Carleton Allen, who counterposed an Austinian omnipotent authority placed high above society, and issuing its orders *downwards*, to situations where law is spontaneous, growing *upwards*, independently of any dominant will.[26] In addition, from a conventional legal standpoint, the notion that cyberspace tribes are potentially lawmaking and law-enforcing places was seen as potentially very useful. The alternative would be to follow all of the laws of all the states that have a plausible claim to make the rules for the people with whom one may be interacting online.[27] The fact is that the Internet is a large and complex legal system that lies outside of all other legal authorities.[28]

Legal autonomy requires that a governance system must be complete, that is, it must possess the full range of powers defined by Raz. It involves the availability of coercive power to enforce group decisions, and a contractual framework expressing the norms, procedures and institutional competencies of participants.[29] Decentralised law would take the form of *customary law* emerging spontaneously from human interaction. This means that law would be first enforced through group norms, the observance of accepted standards of behaviour. Religious orders have long had their own rules outside of sovereign government, forming complete governance systems. They articulate norms and institutionalise and coerce compliance by the prospect of expulsion from membership and from religious grace.[30] More generally, virtually every citizen of a modern state is a member of private organisations: political, professional and trade associations, country clubs, national fraternities and non-profit organisations such as churches, athletic leagues, Boy Scouts.[31] All of these groups exercise some powers of self-governance. Legal scholars have also examined social groups that resolve disputes outside the legal system, such as cattle ranchers and professors at academic research institutions.[32] Members of these close-knit groups develop and maintain norms so as to maximise the welfare that members obtain in their everyday affairs with one another.[33] For example, it was observed that owners of livestock bear responsibility for conduct of their animals, and that norms are enforced thanks to self-help such as phone calls to owners of trespassing cattle, gossip, subtle threats and mild retaliation.[34] Elinor Ostrom observes that 'increasing the authority of individuals to devise their own rules' results in processes that allow social norms to evolve and thereby augment the probability of reaching better resolutions of collective action problems.[35]

What distinguishes online tribes from other self-governed offline groups is the constant presence of disruption and aggression. Archaic force can manifest itself in a purely disruptive and attention-seeking manner, for example in the inverted prestige of 'trolls' who unite all against them. Trolls typically poison discussions by engaging in strategies such as gratuitous abuse, changing the topic, focusing on form rather than substance, emphasising affective issues, claiming that they are being persecuted, and the like. The challenge for a full understanding of online authority is to account for all the dimensions of archaic force, including its contribution to self-governance. As an illustration, we can take the example of 'flaming'.

Flaming has been defined by communication researchers as hostile or incendiary messages whose meaning varies with the intention of the sender, the interpretation of the receiver and that of an outside observer.[36] What may appear insulting to an outside observer may in fact be a message which both sender and receiver understand as sarcastic or humorous. 'True flames' are therefore 'purposeful negative violation of interactional norms', where the message is intended to be offensive and is perceived as such by receivers and outside observers.[37] In this scenario, flaming operates solely at the inter-individual level, whereas it also has a collective significance. In a self-regulatory context, flaming can operate as a form of discipline, such as when newbies are flamed for asking questions which have previously been answered. Moreover, flaming, and the recognition that it is occurring, which may lead to objections by third parties who witness the flame, serve to codify, reaffirm or contest norms.[38] Authorising oneself to address the violation of a norm constitutes the basic building block of communicational online authority – the invocation of a tribal rule to correct others' poor grasp of communal standards. There are striking similarities with the behaviour of children who use arguments to create a micro-social hierarchy in which the successful accuser is seen to be socially higher than the accused rule breaker.[39]

Archaic force plays an important role in motivating the emergence of norms. Benatouil points out that pragmatic sociology would benefit from a social, political and historical recontextualisation of regimes of action: which ones are valued over others? When is no justification required?[40] In other words, just as critical sociology needs to take into account the reflexive capacities and sense of justice of 'ordinary people', pragmatic sociology would do well to demonstrate a more critical understanding of the historical production of disadvantage. This conceptual integration can be illustrated by the example of online tribes, where there subsists, underneath the sovereign and charismatic orders, a zone of rude aggression which is primarily the site of ritualised male proofs of valour and honour. This *archaic residue* can be challenged as illegitimate or obsolete, but it also tends to manifest itself in more insidious ways. As was noted in the previous chapter, archaic inequality can appear as the seemingly natural advantage of male incumbents (online index charisma), or as the unproblematic superiority of male possessors of computer engineering prestige (online hacker charisma). Table 1 (page 80) summarises the properties of online authority regimes.

My analysis does not subscribe to pragmatic sociology's radical abandonment of groupings or collective being, where the public is seen as a phenomenon that leads beyond proximity, towards others treated in a general way. Activity of the sort dealt with here is of necessity a collective affair, as it can only occur within, and derives its meaning from, a collective project. The affective dimension of working in concert and of being validated plays a central motivating role.

Table 1 Regimes of Online Authority

	hacker-charisma authority	index-charisma authority	sovereign authority	archaic force
role	elder, integrator	hub, bridge	judge, enforcer	flamer, troll
act	compile	connect	banish	confront
justification	epistemic	topological	procedural	honour
object	explanations	rankings	deliberations	abuse
space	project	network	assembly	forum

It is beyond the scope of this book to examine comprehensively every possible logic of online action: actors have plural identities, and the registers of economic profit or love, for example, will not be dealt with here. Within this diversity, I will only look for regularities in governance and justice in the context of epistemic tribes. The activities of authority figures in early Internet autonomous projects offer important insights in this respect. As integrators, they had to attract contributors and maintain project integrity and quality; I now address more precisely the logics of autonomous policing and adjudicating, and in particular the sanctions used to enforce norms and laws.

Norm Enforcement: Netiquette and Wizocracy

Online power is manifested by technically controlling or disrupting the communications of other users.[41] Early Internet researchers considered offenders as indulging in a form of pathological behaviour, to the extent that some declared that 'sociopathy has been a major part of our virtual interaction from the beginning'.[42] The guilty parties soon emerged – enter the SNERTs ('snot-nosed Eros-ridden teenagers'). Treatises were produced to advise system administrators on how to deal with their disruptive activities. On MOOs, establishing standards would make it easier for wizards to manage avatars uniformly and fairly, making policing less open to the vagaries of individual judgment. However, this solution generated its own problems: how should the rules be interpreted?[43] In theory, it is easy to enforce rules in electronic social networks.

When the attachment between people is tribal, based on direct relationships, threat of exclusion from the network may be a powerful enough incentive to induce compliance with the rules.[44] Norms develop best in a small and static community because they derive whatever legitimacy they have from group endorsement, and because the internalisation of rules takes time.[45] Members must think alike, share a history, and make sanctions known to new members. Sanctions are also easier to enforce in smaller settings, particularly if members must act collectively.

'Netiquette' was an attempt to devise norms that would be applicable everywhere on the Internet, regardless of the type of community. Usenet, the conferencing system in which posts could be copied from newsgroup to newsgroup, was the primary site of the development of netiquette. Its primary aim was to socialise new entrants who were not familiar with the conventions of electronic communication. Different iterations of netiquette were posted on newsgroups, in FAQs and later in books. They covered a wide variety of issues, from advice, such as the proper use of emoticons, to severe warnings against unacceptable behaviour. For example, an Internet Engineering Task Force (IETF) version of netiquette forbade the use of chain letters in no uncertain terms: 'Never send chain letters via electronic mail. Chain letters are forbidden on the Internet. Your network privileges will be revoked.'[46] In the 1990s, long before mass broadband and the storage of gigabytes of data on key-ring devices, data storage and bandwidth were important issues. Netiquette guidelines reflected this concern. Since, unlike in the case of the post office, the telephone or broadcast media, the cost of delivering electronic communication is borne equally by sender and recipient, strict rules governed both the size of '.sig' files that identify speakers (so as not to waste bandwidth) as well as the cross-posting of irrelevant message to many groups.

Other rules, not directly linked to technological capacity, are still in use in many email lists. For example, great care should be taken when editing others' words when replying to posts so as not to distort or obscure meaning or authorship; new entrants should read mailing lists and newsgroups for one to two months before contributing so as to understand the local culture and customs; it is improper to mix personal and public information (as when one replies to all members of a group or list instead of just one); people should check a group's FAQ before posting questions; they should also request that answers to queries be sent by private email, not to the group, and offer to write a summary of all the responses received. In general messages were to be brief and to the point. As for pointing out others' errors in typing or spelling, the IETF guide did not mince words: 'more than any other behaviour [these] mark you as an immature beginner'.[47] Finally, commercial use of common space was banned: 'In general it is considered nothing less than criminal to advertise off-topic products.'[48]

Most of these guidelines were uncontroversially accepted by Usenet contributors. The question of how to enforce them, however, met with less consensus. On the one hand, the rules usually specified that if someone breached netiquette, they should be politely told by private email, or as the popular phrase had it, 'don't bite the newbies' (this is now a core Wikipedia precept). On the other hand, the male hacker world which informed much Usenet behaviour condoned, to a certain extent, the use of flaming as a social sanction.[49] For example, people posting to a newsgroup requests for information which was widely known to be available elsewhere would, more likely than not, receive suggestions to 'RTFM!' (Read The Fucking Manual).

The common-law approach was most severely tested when dealing with clearly antisocial behaviour. Though 'trolling' originated on Usenet, the same phenomenon has also occurred on lists, where the same basic problem complicated policing: consensus is hard to achieve online. By all accounts, the most effective way of dealing with provocateurs is shunning them, rather than confronting them, which only makes them happy, or banning them, as they will invariably return. However, effective shunning of a disruptive individual requires a group consensus to follow through on ignoring the individual. In Herring et al.'s observation of attempts to deal with a highly disruptive troll, despite widespread agreement that ignoring 'Kent' was a good idea, many participants continued to argue with him, undermining the group's attempt to shun him.[50] The belief in the universality of the social contract leads to a blindness to the attention-seeking nature of trolls. Herring et al. suggest that centralising authority through a moderator or admin is a necessity; further, complete consensus for the removal of a troll should not be mandatory.[51]

When bandwidth was precious, a crime rivalling trolling was *spamming*, the clogging up of the network with irrelevant or irritating information. Spamming on Usenet referred not so much to the length or content of a message as to its cross-posting to multiple lists, so that users received the same information many times. This waste of disk space officially became the worst form of antisocial pollution on 12 April 1994, when lawyers Laurence Canter and Martha Siegel posted an advertisement for legal services on thousands of newsgroups. The angry response this generated went beyond flaming (though flames were plentiful). A newsgroup devoted specifically to dealing with Usenet abuse, alt.current-events.net-abuse, was established on 25 April 1994. Then self-styled vigilantes emerged to attack spammers with 'cancelbots' – programs that systematically removed postings from sources identified with spam. As might be expected, reactions to these cybersheriffs on the electronic frontier (in the Libertarian parlance of the era) were mixed. For Benjamin Wittes, the actions of Cancelmoose, the most famous of these enforcers, were justified, because this individual took pains to avoid going beyond the community consensus on which cancels were legitimate. Cancelmoose posted a justification for each of his cancels and included in that justification 'a copy of the original message and a list of the newsgroups from which the spam was removed [as well as] information allowing sysops to override the cancels'. Cancelmoose also took pains to emphasise that the content of messages did not influence his decision whether or not to cancel them.[52] This last point is important, as the last thing Cancelmoose wanted was to be seen as an opinionated censor rather than a fair enforcer of the accepted rules of Usenet. Censorship contradicts the autonomy principle and is therefore perceived as intolerable.[53] This is summary tribal law, dispensing with liberal democratic defences against injustice such as due process and the rights of the accused. There was no recourse. If one was sentenced to the 'Usenet death penalty' (all posts originating from a site were cancelled or not forwarded) that was *it*. The death penalty was administered by small group of hackers who called themselves the Subgenius Police Usenet Tactical Unit Mobile or SPUTUM.[54]

For Mark Lemley, this was the worse kind of vigilantism: norms of Net behaviour were not being decided by the average user or by consensus; rather, a small group of individuals armed with technical weapons imposed social sanctions as they saw fit.[55] By contrast, in Perritt's view Cancelmoose acted in accordance with the consensus of the participants in the Newsgroup; this consensus was equivalent to the combination of a jury verdict and a warrant: Cancelmoose acted like a deputy sheriff executing an arrest warrant after a criminal conviction. This was fair, though sheriffs are usually not self-appointed. Perritt did concede that the consensus authorising action was much more fluid and informal than a normal jury verdict and might therefore lead to unfairness.[56]

After Usenet, another classic example of self-regulation in the early Internet was the emergence of law on MUDs and MOOs where people used 'avatars' to communicate and interact, foreshadowing today's 3-D virtual environments, such as Second Life. A famous example was LambdaMOO, a popular text-based virtual environment launched by Pavel Curtis and others from Xerox Labs. LamdaMOO is particularly interesting from the perspective of online authority because it represents a clear example of the passage from one regime of online authority to another. In general, MUDs and MOOs were controlled by 'wizocracies'. On LambdaMOO, wizards were responsible for both technical integrity and social control: they made the rules, decided when to increase player quotas (the quantity of disk space reserved for the objects and spaces they created), attempted to resolve disputes, and sometimes destroyed 'incorrigibly antisocial' players.[57] They were accordingly treated with exaggerated deference by other users. A well-documented case of 'virtual rape' on 9 December 1992 convulsed the MUD.[58] Following this incident, Hakkon (aka Pavel Curtis), LambdaMOO's arch-wizard, decided to abdicate his authority. He wrote:

I believe that there is no longer a place here for wizard-mothers, guarding the nest and trying to discipline the chicks for their own good. It is time for the wizards to give up on the 'mother' role and to begin relating to this society as a group of adults with independent motivations and goals. So, as the last social decision we make for you, and whether or not you independent adults wish it, the wizards are pulling out of the discipline/manners/arbitration business; we're handing the burden and freedom of that role to the society at large.[59]

How does a group constitute itself as a sovereign collective? In political-philosophy terms, a group of strangers would only accept as legitimate laws which they would agree to enact as autonomous legislators and to follow as law-abiding subjects. The means to achieve generalisable laws would be through open and equal communication. The appearance of deliberative procedures is therefore a key sign of the passage to the sovereign form of authority.

On the MOO, wizards were to become mere technicians applying the consensual decisions of the community. To this end, Hakkon instituted a system whereby users could launch petitions to change all aspects of the MUD's operations and use 'LambdaLaw' to resolve issues such as freedom of expression versus protection from harassment.[60] When a petition was created, a mailing list was simultaneously established to discuss it. When a petition garnered ten signatures, its creator submitted it to a wizard who vetted whether it was appropriate or implementable. Vetted petitions obtaining 5 per cent of the average vote count on all ballots or 60 signatures were then transformed into open ballots and required a two-to-one margin to pass. They were kept open for two weeks before expiring.

There remained the question of the accountability of wizards.[61] As a first step, arbitration procedures were established. Anyone could initiate a dispute, but they must have to be deemed to have suffered an actual injury. This was decided by volunteer arbitrators, who were chosen by disputants or otherwise randomly assigned to a case. Parties could not initiate a dispute against more than one person nor initiate two disputes simultaneously.

Mailing lists were established for every dispute. Arbitrators could advocate almost any form of punishment within the MOO, such as reductions in quotas of allowed activities, and recycling (destruction) or toading (banishment) of avatars. Adjudication and enforcement mechanisms, such as the creation of a discussion space for each dispute, and a system of graduated sanctions, are a staple of mass online tribes such as Wikipedia. In general terms, LambdaLaw was not dissimilar to its inspiration, the American legal system. This can be seen in the mechanisms justifying the sovereign authority of arbitrators (such as process-based systems for determining the legitimacy of decisions); in an individualistic conception of property rights applied to virtual objects; and in the name of proposed institutions, which reflected their originating legal culture, such as the Lambda Supreme Court and the Lambda Bill of Rights.[62] Participants also took it for granted that free speech was guaranteed, though no such provision existed.

Of particular import was the *dialogic* nature of the arbitration process, which was encouraged through the creation of specific mailing lists. No judicial distance was maintained, and MOO arbitrators frequently submitted drafts of their decisions to gauge community support. Once again, similar arbitration mechanisms have been adopted in tribal systems such as Wikipedia and Debian. LambdaLaw also anticipates the question, so frequently raised in Wikipedia, of the capacity of a community, when private information is involved, to make an informed decision without having all the facts at its disposal.

The LambdaLaw process had one serious limitation: arbitrators could take decisions affecting only the two parties. No punishment could limit the rights of other players or call for new law as a result of arbitration. Jennifer Mnookin notes that this lack of precedent signified there was 'no guarantee that similarly situated disputants would be treated in the same manner'.[63] The MOO also suffered from the absence of an authoritative body for resolving interpretative differences. The establishment of a Judicial Review Board was proposed, so as to avoid revisiting similar arguments over and again. Despite strong support, this proposal did not manage to achieve a two-thirds majority. As a footnote to the LambdaMOO story, it should be mentioned that wizardly fiat was reintroduced in May 1996, as the only way of dealing effectively with disruptive players. Wizards could not limit themselves solely to technical decisions, as they had to administer the server's security procedures as well as secret correspondence between real-life identities and MOO identities.

Since the mid-to-late 1990s and the democratisation of Internet access, non-technically able users have joined in online information sharing and cooperative work. In today's Web 2.0 environment, the range of those who exercise technical control has dramatically risen. Examples dealt with in this book include Wikipedia articles and progressive political weblogs.

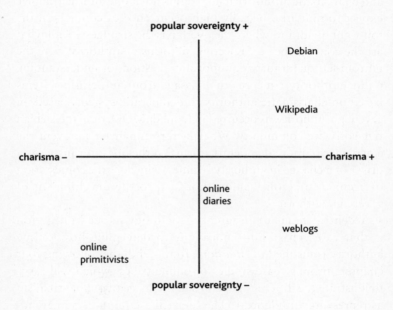

Figure 1 The space of online tribal authority

Figure 1 presents a spatial representation of the field of online authority in the form of four quadrants intersected by two main axes of charisma and sovereignty. Online tribes are positioned in the quadrants according to the structure of their authority relations. The top-left quadrant is empty as it corresponds to the space of governmentality: that is, popular sovereignty with no autonomy. An example would be the online networks developed by political campaigns to give their supporters the impression that autonomous grassroots activism is being fostered.

Tracking Authority in Four Online Tribes

The schematic representation of the space suggests that online authority regimes do not exist in a perfectly pure state, but rather in various forms of combinatory amalgamations. Since sovereign and charismatic authority tend to contradict one another, can we expect to find an especially high occurrence of conflicts in the top-right quadrant, where the overlap is the strongest? And what are the consequences of rejecting all forms of authority, as in the bottom-left quadrant? An important question for understanding online authority concerns the passage from autocratic systems based on charismatic authority to democratic systems based on sovereign authority, as occurred in LambdaMOO, or in the Debian free-software community. What factors enhance or prevent the emergence of more democratic forms? If the quest for justice is a driver of a change in authority regimes, why does the contestation of archaic force on weblogs not lead to the development of sovereign authority?

A central tenet of pragmatic sociology's focus on action is that research should endeavour to determine how online authority is justified *in situation*. To this end, the following four chapters propose case studies of online tribes, which will serve to illustrate the different positions in the space of online tribalism as represented in Figure 1. The four tribes are: radical anarcho-primitivist websites and forums (lower-left quadrant); progressive political weblogs (lower-right quadrant); the Debian free-software mailing lists (upper-right quadrant); and the English Wikipedia wiki and mailing lists (upper-right quadrant). As mentioned above, the upper-left quadrant is tribeless. The analysis will compare the four tribes in terms of three main parameters: project and space;

authority structure; and conflicts and enemies. These categories are briefly presented in the remainder of this chapter.

Project and space. How does the project embody autonomy? What kind of computer-mediated communication is it, and what technical possibilities are available for ordinary users? In terms of participation, how does recruitment operate, and what kind of contract is offered to new entrants? What are their roles, duties and privileges? And how are boundaries maintained?

Authority Structure. What is the relationship between expertise (learned authority) and leadership (administrative authority)? To what extent is charismatic or sovereign executive authority distributed? What are the tools of governance, such as norms and rules, monitoring, adjudication, and enforcement mechanisms? Is authority strong, weak or inconsistent?

Conflicts and enemies. Antagonism is central to tribal activity. Rules regulate the integration of patches or the positions of people. Conflicts, triggered by the application, justification or absence of rules, are the means by which people affirm their adherence to, or rejection of, the rules and the authority order which they underpin. A central concern will be the role of *enemies*. The definition of outside enemies is vital to coalesce project cohesion and exclusionary boundaries. This is all the more the case when projects experience internal conflict, as outside enemies will help to reinforce project solidarity.

PART II

5

THE LAST ONLINE TRIBE:
primitivism.com

In its broadest terms the contrast between tribe and civilisation is between war and peace. In the social condition of Warre (Hobbes), force is a resort legitimately available to all men.

Marshall D. Sahlins, *Tribesmen*

This chapter focuses on the lower-left quadrant, which has no charismatic or sovereign authority. The data repository primitivism.com is an online project that explicitly rejects any notion of networked interactivity and remixing. Since this sector of the radical underground does not take part in the link economy, communication and conflict is organised around texts, particularly disputes about the interpretation of theories and historical events.

Project: Web 0.0

Primitivism.com defines primitivism as the pursuit of ways of life running counter to the development of technology.[1] Online-primitivists advocate the end of all forms of domination, yet do not embrace *online* autonomy. They exemplify the tension between means and end, capitalist media and radical message, which characterises many far-left propagandists on the Internet. Beyond technology, it is civilisation itself which they reject. Anarchists had traditionally critiqued the manifestations of hierarchical thinking and authoritarian social relations; anarchists embracing primitivism, or anarcho-primitivists, attack the assumptions behind that thinking. They abhor techno-industrial development, which they equate with individual oppression and environmental destruction. On the biospheric level, they paint an apocalyptic portrait of species extinction, proliferating dead zones, the pervasive poisoning of air, water and soil. In terms of individual autonomy, they argue that we live in a world in which the accumulation of technical knowledge is astonishing, and yet we are probably much more lacking in technical know-how than our ancestors: technology can only be created and repaired by *someone else*. For Theodore Kaczynski (the so-called 'Unabomber') the freedoms we have are those consistent with the system's ends, such as the economic freedom to consume, or press freedom to criticise inefficiency and corruption; however individuals or groups are devoid of the true power to control the life-and-death issues of existence – food, clothing, shelter and defence.[2] Anarcho-primitivists advocate a return to a tribal mode of living, based on small-scale sustainable communities of hunters and gatherers or permaculture-practising farmers.

In Bourdieuan terms, online-primitivism is a highly autonomous political–cultural field of restricted production (oriented towards other producers), in contrast to fields of large-scale production (produced for general audiences). As in all marginal fields, the production and exchange of cultural artefacts represents the means for people to engage in the genuinely felt rejection of dominant values whilst also exhibiting their underground distinction from mainstream, or 'common' culture.[3] Learned authority on the online primitivist field derives primarily from the provision of specialised knowledge: individuals and groups adapt offline magazines to the Web (*Green Anarchy, Do or Die, Green Anarchist*), constitute online archives of theory (Primitivism, Insurgent Desire), offer practical advice about primitive techniques and 'rewilding' (Wildroots, Earth Skills, Abotech, Slinging) or set up magazine and book distribution hubs (Beating Hearts Press, Re-Pressed). Aside from these data repositories anarcho-primitivism exists online in occasional discussions in anarchy-oriented discussion boards.

What distinguishes the primitivist online corner of the Net from other fringe, radical or 'extremist' fields, what gives it its special flavour is, of course, its inherently contradictory nature. Nowhere is the paradox more apparent than in the name primitivism.com, and I accordingly focus on this site in order to answer the following questions: What does digital media built by technophobes look like? And what forms of authority can be detected there? The principal purpose of primitivism.com is to propagate the theoretical underpinnings of a world view; it is not an activist project. To this end, the website offers a selection of interviews, articles and book extracts organised in categories such as primitivism, technology, health, anthropology, politics, etc. All the texts have been reformatted and appear in front of a similar background, with no bibliographical information such as dates or publishers, reinforcing the impression that they constitute a homogeneous whole.

The site's front page has been frozen since 23 November 2002, which is the date of the last update indicated in the 'News' section. That the front page has not changed since then can be ascertained by checking the Internet Archive's Wayback Machine.[4] The site is not dead or inactive: the Wayback Machine indicates that changes have been made to the site (none were made in 2003, 2004 and 2005), though the Wayback Machine does not indicate what these changes might be. It bears repeating: *the front page of an active website has not been modified in over five years*. This stasis contradicts the very essence of the World Wide Web, the fact that it can be endlessly rewritten and updated.

The website's insularity constitutes another striking difference from dominant Internet practice, flowing on from online primitivists' deep ambivalence about the use of their medium. On the one hand, by posting large quantities of free data, the website is an active participant in the Internet's decentralised 'gift economy'. But at the same time, primitivism.com purposefully does not participate in the Internet 'link economy', the attribution and seeking out of links. For Fuchs, the defining operation of 'Web 1.0.' is that it is a self-referential medium: when a new link is created, the system refers to itself by actualising its content.[5] In contrast, primitivism.com *never* offers a link to other sites or online repositories featuring information on the radical books and articles it presents.[6] As for acknowledging links *to* primitivism. com, the 'News' section mentions articles published in *Reason*, *The American Prospect* and *Zmag* which reference the site, dated from December 2001 – and that is all. More active and activist-oriented primitivist websites such as Green Anarchy mostly link to organisations supporting prisoners. The online primitivist project can properly be called 'Web 0.0', because their online presence constitutes its own negation: networked primitivists do not hyperlink, or exclusively link to offline 'sites' such as prisoners. They are only interested in what lies outside the network since, ideally, they would live in an unconnected world.

Refusing to use standardised blogging tools in favour of web design signifies that more technical sophistication is required: blogs automatically archive content, websites need to be designed. For all their refusal of domination, primitivist websites maintain a tight control over their internal communication. There is a rigid separation between content provision on websites and audience interaction and feedback on bulletin boards and forums such as flag.blackened.net, anti-state.com and www.anarchymag.org/tracker.

Authority: The One Inside

Though the author of the primitivism.com site may not seek out a high profile, a Google search for 'primitivism' will return the site's front page and its article "What is primitivism" in the top four results, alongside the Wikipedia pages on primitivism and anarcho-primitivism. A completely autonomous form of expertise, whereby a superactivist enclave was established with no outside input, has generated high index authority. Who then are the experts?

Aside from the site's creator, John Filiss (who has seven contributions), the two most prolific authors on the website are John Zerzan (eleven contributions) and Bob Black (nine contributions). These are the site's anti-authorities. John Zerzan, a central figure in the *Green Anarchy* magazine, is the most well-known primitivist author. Zerzan asserts that 'mounting evidence' indicates that before the Neolithic shift from a foraging or hunter–gatherer mode of existence to an agricultural lifestyle, most people had ample free time, considerable gender autonomy or equality, an ethos of egalitarianism and sharing, and no organised violence.[7] Zerzan believes the root cause of the problem to be civilisation, that is to say the domestication of plants, animals and humans which led to patriarchy and the division of labour.[8] Bob Black has a legal background and has been active in the anarchist and underground scenes, where his caustic humour and provocative style have fuelled many controversies, since the late 1970s. Initially influenced by the Situationists, he later developed an interest in primitivism and anthropology. Both Zerzan and Black have published several books.

Primitivism lies at the confluence of several strands of radical thought, most of which are represented on primitivism.com. There is first the anti-technological anarchism of Detroit's *Fifth Estate* journal, in which Fredy Perlman first wrote of the 'song and dance of primitive communities', and David Watson extolled the virtues of pre-industrial systems and tribal religions.[9] A notable exponent of this view is Theodore Kaczynski, but, as mentioned previously, his manifesto is not featured. A more mainstream type of anti-technological criticism includes Jacques Ellul, Langdon Winner and Kirkpatrick Sale.[10] The site also includes extracts from books by a diverse group of authors such as Adorno, Gurdjieff, Freud, Ivan Illich, Bill Joy and David Attenborough, to name just a few.

Radical environmental groups engaged in direct action are not represented on primitivism.com. The Animal Liberation Front and Earth First! were inspired by Edward Abbey's popular tale of eco-sabotage[11] and by the 'deep ecology' of Arne Naess[12] and of Bill Devall and George Sessions;[13] Earth First! believes humanity is in the midst of 'an unprecedented, anthropogenic extinction crisis'.[14] Another omitted strand is eco-feminism, which describes commonalities between the subordination of women, indigenous people and nature in terms of their inferior positioning in Western thought and their common exploitation by the capitalist economy.[15] Though figures such as Black and Zerzan are certainly influential, they can hardly be described as charismatic. In the primitivist sphere, that epithet should be reserved for exceptional individuals whose connection with transcendence results from their writing being given extraordinary weight by their actions. It is fitting that the most famous primitivist anti-leader lived a life of solitude, first in the woods of Montana, and now behind bars. Kasczynski, sociopath and killer, whose name was removed from the list of authors of primitivism.com, is the one inside, the absent figure at the centre of the primitivist pantheon.

Conflict: The Bookchin Brouhaha

In the offline world, primitivists, green anarchists, animal liberationists and radical environmentalists are sometimes involved in violent protest, conflict and sabotage. Several activists in Oregon were sentenced in 2007 to long prison sentences for several acts of eco-sabotage against targets such as a horse slaughterhouse, an SUV dealership, a scientific research centre, logging companies and a ski resort, because the charges were classified as domestic terrorism. In what has been dubbed a 'green scare' campaign, the FBI has asserted that groups such as the Earth and Animal Liberation Fronts constitute the 'No. 1 domestic terrorism threat'.[16] If anarcho-primitivists were to engage in similar acts of resistance, sabotage or protest *online*, they would be faced with the problematic notion of having to master the technological tools of 'hacktivism', so as to conduct (for example) website defacements, or computer virus infestations, or 'distributed denial of service' attacks, whereby a targeted website is inundated with browser 'hits', causing it to crash. They would have to become *primitivist hackers*, taking the contradiction to dizzying heights.

The technophobia which makes online primitivists poor linkers also makes them poor attackers outside their tribe, as in this field conflict takes the form of textual exchanges, and online primitivists do not wish to converse with the state and corporations. This means that the erection of boundaries occurs solely within the tribe, so that online primitivist practice is the precise opposite of Sahlins' characterisation of peacemaking as 'the wisdom of tribal institutions [because] in a situation of Warre, where every man is empowered to proceed against every man, peacemaking cannot be an occasional inter-tribal event. It becomes a continuous process, going on within society itself.'[17] This state of affairs is possible, of course, because the conflicts occurring on the online primitivist field are symbolic fights involving no loss of life, limb or liberty. Because primitivist learned authority exclusively stems from its possessor's familiarity with offline thought, there can be no recognition on primitivism.com, and on the primitivist Internet in general, that online forms of authority exist. It is impossible to distribute something which one refuses, so that in this most autonomous of areas, there is no decentralisation of legitimate power, be it charismatic or sovereign. There is no patch to integrate, no post to link to, no entry to correct. And since administrative authority is never in play, learned authority becomes, by default, the pivotal stake of debate and conflict.

Conflicts do not apply to people but to texts: the only integration of another's work is by 'fisking' it (making aggressive comments). The central dividing axis of the contemporary ultra-left field is the question of what is being opposed. Among this close-knit group of enemies, theoretical and ideological chasms hinge upon whether society is defined as 'modern' (dispensing with traditions in the name of progress), 'industrial' (filling an artificial world with technical objects, abolishing nature and humanity), 'capitalist' (subsuming everything to the commodity) or 'spectacular' (negating true life). Primitivists reject all forms of conventional progressive or radical politics because they decry them as embracing the Enlightenment-born ideology of progress and techno-scientific reason. Enemies are, by definition, all those who do not adhere to anti-civilisation views.

Primitivists eventually found themselves at odds with a prominent voice on the anarchist field, that of Murray Bookchin. In *Post-Scarcity Anarchism* (1971) Bookchin had argued that anarchism represented the application of ecological ideas to society, based on empowering individuals and communities, decentralising power and increasing diversity.[18] In 1987 Bookchin published a pamphlet criticising deep ecology's nativism, asserting that there were 'barely disguised racists, survivalists, macho Daniel Boones, and outright social reactionaries' employing the term ecology to express their views.[19] He also described the deep ecologist goal of reducing the world's population as an act of 'eco-brutalism' reminiscent of Hitler.[20]

A few years later, Bookchin launched a broadside against what he called 'lifestyle anarchists'.[21] Contrasting the personalistic commitment to individual autonomy of 'lifestylers' with his collectivist commitment to social freedom, Bookchin directed some of his most abrasive comments towards primitivists, describing their 'edenic glorification of prehistory' as 'absurd balderdash'[22] and ridiculing Zerzan's 'reductionist and simplistic' notion that self-domestication through language, ritual and art inspired the subsequent taming of plants and animals.[23] This generated a flurry of angry anarcho-primitivist responses, notably from Bob Black and the *Fifth Estate*'s David Watson.[24] Bookchin responded to these critics with a new essay in which he reprimanded Watson for serving up 'all the puerile rubbish about aboriginal lifeways [of the 1960s]'[25]; he concluded his piece by dismissing Black's 'irresponsible, juvenile bravado'.[26] Bob Black shot back with a rant in which he derided Bookchin's 'reiteration of the bourgeois Hobbesian myth of the lives of pre-urban anarchist foragers as solitary, poor, nasty, brutish and short, in dramatic contrast to the life of Murray Bookchin: nasty, brutish, and long'.[27] In addition, since Bookchin had based his criticism of the primitive affluence thesis on a book by the anthropologist Edwin Wilmsen which affirmed that the !Kung Bushmen – contrary to primitivist orthodoxy – lived miserable lives,[28] Black lost no time in attempting to demolish Wilmsen's credentials in yet another essay.[29]

This polemical back-and-forth took a more concentrated form during the exchange between Ken Knabb, the main translator of Situationist texts in the United States, and primitivism.com creator John Filiss. Knabb's essay on revolutionary politics, *Public Secrets*, included a critique of anarcho-primitivism from an ultra-left perspective. Knabb affirmed that primitivism offers no practical means of achieving its goals in a libertarian manner, because a mere revolution could never be enough to satisfy the 'eternal ontological rebelliousness' of primitivists.[30] This section of Knabb's text was posted by John Filiss on the 'Anarchy Board' discussion list together with Filiss's point-by-point comments and refutations. Knabb then reappropriated this cut-up and added his comments to the mix, posting the result on his Bureau of Public Secrets website.[31] Not to be outdone, Filiss then copied this new version, added yet another layer of commentary – commenting on Knabb's reactions to his earlier comments to the original text – and posted it on primitivism.com, under his own signature.[32] The combination of text editing and digital networking technologies facilitates a synchronous presentation of that most diachronous form of intellectual exchange, the literary dogfight, in which authors can seize their opponents' text, modify it by changing the font characteristics or by adding breaks, and insert their own thoughts, in a potentially never-ending conflictual palimpsest. In a setting where authority is seen as illegitimate, there is little formalisation of norms, and no appeasement mechanisms. The only rules are that anarchist principles, such as always refraining from coercing anyone, should be respected. Since there is no supreme authority to adjudicate disputes, no resolution occurs, and conflicts can continue for a very long time. The historical records of past conflicts archived on primitivism.com do not reach the level of verbal violence found in anarchist forums, where 'flaming' is an accepted part of discourse.[33]

A widespread means of disparaging an opponent's views is to use a derogatory term such as 'leftist', 'authoritarian' or 'trendy'. This last term was used by anti-primitivists such as Bookchin, who reviled 'today's fashionable technophobia',[34] and Knabb, who asserted that he conducted a brief debunking of 'trendy technophobia'.[35] Conversely, it was also levelled at 'insurrectionary anarchists' by primitivist elder Zerzan, who described this rival tribe as a 'trendy and possibly hollow movement'.[36] Knabb accused Filiss of summarily ejecting someone from his forum, an inherently authoritarian act.[37]

Ritualistic arguments over earlier examples of ideal social organisation are also used as weapons. As previously noted, anarcho-primitivists base their claims on anthropologies of hunter-gatherers. Their 'leftist' enemies hark back to examples of popular uprisings such as the organisation of society by anarchists during the Spanish Civil War, 'probably the single richest example of the potentials of autonomous popular creativity'.[38] Yet frequent references to history and anthropology present primitivists with a problem: how to refer to 'experts' without appearing to embrace a conventional system of hierarchical knowledge. In the conflict against Bookchin, Bob Black mocked his opponent's stuffy academicism: Bookchin, wrote Black, is a 'self-important, pompous ass'.[39] But he also reproached his lack of scientific rigour: 'Unlike in [another book] Bookchin this time provides a source for his claim that ... '[40] 'In the sequence in which Bookchin places it, the Feyerabend quotation – unreferenced – looks like a summons to freak out.'[41] This ambiguous relationship to high culture and scientific standards is typical of underground culture in general and of anarchists in particular, who wish to be perceived as authoritative, but not authoritarian.

The ideological rigidity of online radicals means there is little attempt to meet half-way or self-deprecate; exchanges tend to be very aggressive and conflicts are rife. Conflict enables protagonists to refine and reiterate fixed positions, but there can be no evolution, as people's identity is inextricably linked to their strongly held position – backing down, changing one's point of view, would be the same as abdicating one's right to exist. There is no place for deliberation in this space. If one considers the frequent use of abusive language (as a synechdoche for the rejection of bourgeois, conventional notions of propriety and property) and the fact that disputes rage on unabated, it may be appropriate to ask whether this behaviour represents, in digital form, a 'state of nature' in which all are at war against all. Online primitivism's incessant production of centrifugal conflictuality may be worth pondering for other radicals, as the decrease in the physical and psychological costs of engaging in internecine warfare (to such an extent that war subsumes all other intent, becoming an out-of-control force) does not necessarily augur well for the capacity of online radical tribes to effect change.

Radical Dissent and the Net

If the primitivist project were successful there would be less and less technology, less and less online primitivism. In the online primitivist field there can be no question of efficiency evaluation, of success in developing projects and in attracting participants. No authority means no distribution. If that were not enough, purification campaigns and the excommunication of antagonists signify that recruitment can never be anything but low.

Radical environmental outrage reminds us that what industrial production has done, is doing, and is *planning to do* to our world should not be blindly assented to, or go unchallenged. But primitivism goes much further and in so doing forsakes any claim to conventional notions of moral legitimacy. From an ethical perspective there is no justification for a belief system which holds that the population of the earth should somehow 'change' from six billion to one hundred million. This is not a philosophy that upholds the common good – or rather, here the common good's application is strictly tribal, applying only to those who deserve to be saved, rather than universal. In this sense the structural position of primitivists in relation to mainstream sectors of Net activity mirrors that of so-called 'hate sites'. For primitivist, far-right and radical Islamist groupuscules, the parameters of acceptable discourse are immutable. Objectives are clear: the overthrow of civilisation, the dominance of the white race, the establishment of a world caliphate. Enemies are identified: industrial civilisation, non-whites, the West. This structural proximity has even been acknowledged by primitivists themselves. When Theodore Kaczynski listed in his manifesto 'rebels against the system', he wedged 'radical environmentalists' between 'Nazis' and 'militiamen'.[42] The Internet offers primitivists the same advantages as far-right groups who search for a virtual community to compensate for a lack of critical mass in their own town or country.[43] As Castells notes, the network structure of the Internet reproduces exactly the autonomous, spontaneous networking of militia groups, who do not have a definite plan, but share 'a purpose, a feeling, and most of all, an enemy'.[44]

All these groups strive to formulate the strongest critique possible. Scott Lash suggests that only immanent forms of critique are possible in the information society: there is no outside space for transcendental critical reflection; the immediacy of information, globalisation, the erasure of boundaries between human and nonhuman mean that critique must operate within information.[45] Oppositional identities such as primitivism, far-right activism and radical Islamism exist to a certain extent within media space, but they also refer to non-mediated reality. These coherent identities, perceived as deriving from race, religion or nature, are in fact the very opposite of technological media. The difference between primitivism and these other radical projects is that if the primitivist vision were to be realised, there would be no more electronic communication and organisation. This is why they are 'the last online tribe'.

The niggling irony of fierce opponents of industrialism and technology communicating via technological networks is not lost on online primitivists. A range of rationalising discursive strategies have appeared, such as the following justifications for electronic communication by a group of primitivists: John Connor complained that it was disappointing that the 'orthodox' ask 'the impossible of anti-tech critics, demanding they *personally* live free of technology' when technological society denies them the possibility of doing so.[46] Jonathan Slyk was less ambiguous, writing that 'the point is not to run away from society and civilisation – but to destroy it'.[47]

While it is unclear to what extent online networking affects the autonomy of the offline primitivist field, the fact remains that establishing primitivist websites, despite the best efforts of members of the tribe to limit their connectivity and accumulation of index authority, cannot but reinforce existing hierarchies of information and power, which are today based on access to networks.

6

THE PRIMARY WAR: dailykos.com

Any political party that can't cough up anything better than a treacherous brain-damaged old vulture like Hubert Humphrey deserves every beating it gets. They don't hardly make 'em like Hubert anymore – but just to be on the safe side, he should be castrated anyway.

Hunter S. Thompson, *Fear and Loathing: On the Campaign Trail '72*

The idea that consumers of commercial media should be able to 'talk back' and reclaim the means of communication for their own purposes is not new. Science fiction fanzines have been in circulation since the 1930s. Their origins lie in the Letters of Comment (LOCs) sections at the end of short-story magazines which spun off to become stand-alone publications. The punk subculture adopted the fanzine early on, starting with Mark Perry's *Sniffin' Glue* in 1976. In the 1980s punk rock cross-pollinated with other underground networks and this, together with the democratisation of photocopying and desktop publishing, gave rise to the 'zine explosion'. Zines were idiosyncratic amateur publications, usually written by one person, which dealt with anything the author was interested in.[1] Mike Gunderloy and Seth Friedman's 'zine of zines', *Factsheet Five*, was the network hub where people sent their zines to be reviewed and exchanged with other enthusiasts; later still, as zines moved from photocopying to offset printing, they were sold to a widening audience. Zines were 'discovered' by the mass media in the mid to late 1990s.[2] Some editors obtained book deals, others burned out, and the rise of the Internet finished off the rest. The central problem affecting zines, distribution woes, was abolished by digital networking; yet the similarities between the liberatory discourse used about the two media forms are striking: just like blogs today, zines were going to 'take back the media'.

Zines thrived on mutual reviews, whilst the essence of the blog is the link, often to an interesting source in the traditional media. Whereas zinedom constituted a clearly distinct system, blogs seek to influence mainstream discourse. They owe much in this respect to the 'gonzo' journalism of Hunter S. Thompson, whose irreverent criticism of Democratic-Party-machine politicians foreshadowed blog critiques of the centre-right Democratic Leadership Council. For some, 'gonzo' stands for drug-addled and in rather poor taste; but for others, it represented a particularly brave form of truth-telling: engaged journalism at its best. A similar concern animates bloggers. As one of Daily Kos's editors or 'front-pagers' asserted: 'We emerged because the megamedia – the oligopress, the pundithugs, the corporatist whoredom of propaganda – were lying to us, and when they weren't lying, they were omitting the truth.'[3] Calling it as you see it: for bloggers, there lies the greatest difference between their sometimes brutal honesty and the objectivity of the establishment press.

Project: The Democratic Noise Machine

Progressive political blogs in the United States came into being in reaction to the controversial 2000 presidential election and Florida recount. Media Whores Online, Bartcop and Talkingpointsmemo emerged to 'follow the debates and criticise the lazy press coverage of the campaigns'.[4] In 2001 Jerome Armstrong started publishing MyDD, which stood for 'My Due Diligence' (and now for 'My Direct Democracy'). He allowed readers to comment in 2002, and, as the budding support for Howard Dean's run at the Democratic primary coalesced on the site, an ex-soldier named Markos Moulitsas Zúñiga ('Kos') began posting there. The lead-up to the war in Iraq saw the rise of Eschaton (written by Duncan Black, aka Atrios) and Daily Kos, which started on 26 May 2002. Kos's first post began with: 'I am progressive. I am liberal. I make no apologies'.[5] A milestone was reached in 2005 when Armstrong and Moulitsas published *Crashing the*

Gate. This book detailed their contempt for the consultants and lobbyists profiting from the Washington 'gravy train', their belief in the need for Democrats to aggressively challenge Republican themes and figures, and their faith in the democratising power of technology to achieve this purpose.

Many progressive bloggers are ex- or would-be journalists. Their desire for the press to operate as a more effective check on the government stems from this professional background, as well as from their opposition to the Bush presidency, and in particular to the way in which this administration benefited from the traditional media's failure to properly scrutinise the government's rationale for the invasion of Iraq. Bloggers' criticism of politics does not extend to political economy: they cannot or will not address the impact of ownership of media on media content – except when it threatens to interfere with the electoral process, as occurred when they protested against the broadcast of *Stolen Honor*, a documentary which they felt slandered the patriotism of John Kerry.[6] In the main, bloggers are not radical critics of the Fourth Estate as an instrument of industrial capitalism, in the manner of Chomsky, McChesney and Bagdikian.[7] Hence weblog writers do not question the traditional news media's narrow range of topics and sources. Though they criticise the collusion of establishment journalism with establishment political power, they validate this journalism by linking to it, rather than to alternative news sources.[8] Bloggers operate mainly as populist fact-checkers – challenging, refuting and correcting perceived errors.[9] Their relationship to the traditional media can therefore properly be called parasitic, as their existence depends on that of the larger organism on which they feed. Where blogs do challenge the media 'system', like zines, is in allowing 'anyone' to participate. The personal voice can be heard. As Stephen Coleman put it,

> to blog is to declare your presence; to declare to the world that you exist and what it's like to be you; to affirm that your thoughts are at least as worth hearing as anyone else's; to emerge from the spectating audience as a player and maker of meanings.[10]

Well-known early bloggers such as Rebecca Blood stereotypically asserted that not only are blogs reclaiming the means of communication from corporations, but that every kind of weblog 'empowers individuals on many levels'.[11] Beyond the promise of rejecting alienation by transforming the media consumer into a media creator, what makes blogs truly new is their interactivity. Against what Castells describes as the mass media's one-way communication, its extension of mass production and industrial logic into the world of signs,[12] weblogs offer the promise of a participatory culture. Self-expression becomes intertwined with community co-production, as do interpersonal and one-to-many communication.[13] Kos situates blogging's impact within the spectrum of open-source politics, activism and journalism, as 'the aggregation of thousands on behalf of a common cause'.[14] Grassroots online organisation formed the basis of the appeal of Howard Dean's primary campaign in 2003, with the use of the Meetup web tool for forming social groups. The result was 'Deanspace', where an energised base, with a strong sense of ownership and involvement, could talk amongst themselves.[15] The Obama campaign emulated this success in 2007–08, albeit in a more controlled fashion.

Interactivity provides the means to challenge the traditional media's ability to define reality. Progressive bloggers, in common with their conservative opponents (the 'war-bloggers' who emerged following the 11 September 2001 attacks and vociferously supported the invasion of Iraq), wished to challenge the monopoly of traditional media on truth claims, based on the collective expertise of tens of thousands of readers every hour. One personal voice, one author, is more easily approachable than the *New York Times*, and blogging technology makes it easy to respond instantaneously to that voice. As Josh Marshall of Talking Points Memo put it, 'there is some real information out there, some real expertise. If you're not in politics and you know something, you're not going to call David Broder.'[16] The *Washington Post*'s Broder, the so-called 'dean' of North American political journalism, symbolises for many bloggers all that is wrong with professional journalists – their proximity to, and complicity with, political elites.[17]

Besides operating as a more truthful information source, another aim of the progressive blogosphere was to counterbalance the perceived influence of Republican think tanks over the terms of the public debate. Rick Perlstein's *Before the Storm: Barry Goldwater and the Unmaking of the American Consensus*, which details how the Goldwater supporters took control of the GOP in the 1960s and 1970s to inflect the ideological foundations of American discourse towards the right, enjoys canonical status among 'netroots' bloggers. Progressive bloggers believe that the resulting shift of the political centre to the right enabled so-called centrist Democrats to score points by being bipartisan, i.e. distancing themselves from their own party, in the process foreclosing the possibility of real change.[18] But bloggers such as Kos are only interested in this ideological contest inasmuch as it enables the winning of elections. For this is their third, and most important objective: *winning*, securing the electoral victory of Democrats over Republican, and of progressive Democrats over conservative Democrats.

Authority: Drinking From a Fire Hose

Daily Kos comprises all the forms of online authority. The central position that Daily Kos occupies in the progressive blogosphere can be statistically verified: the blog has 150,000 registered users, averages 500,000 visits a day, and regularly reaches the top of the various sites which classify the popularity of weblogs, whether based on hyperlink patterns or citations of posts. The blogosphere is indeed the domain of pure index authority. Whether they concern comments, links or hits, *rankings* are the currency of blogs. Kos benefits from preferential attachment: any new progressive blog links to him. If links are votes, then Kos has consistently been elected king of the blogs. Naturally a link from such a source will drive traffic (and associated advertising revenue) up. Henry Farrell has noted that some high-traffic blogs play a crucial role as clearing houses for attention and information, and that 'their power is one of influence, framing and persuasion rather than hierarchical command'.[19] The ability to direct the attention of hundreds of thousands of people means that leading blogs have the power to confer legitimacy on issues and campaigns.

The structural role of elite blogs is to resolve the tension between the unfulfilled desire of most bloggers to find readers, and the insurmountable task faced by blog readers who want interesting content but are faced with an ocean of offerings. Top-ranked blogs constitute 'a focal point at which bloggers with interesting posts, and potential readers of these posts, can coordinate'.[20] Lesser-known bloggers with interesting information will not be content with simply posting it to their own page; they will usually alert an elite blogger. This is much more cost-efficient for top blogs than having to trawl through thousands of less popular blogs to find interesting material. If the A-lister publishes the information, the originator may receive a hat tip ('h/t') or acknowledgement in return, in addition to the all-important link. Elite blogs then serve as conduits to mainstream media, as reporters can reliably expect that they will serve an aggregative function, distilling from the great mass of content certain central tendencies; and they also serve as portals or conduits for this information.[21] The purpose of autonomous individuals on the progressive blogosphere is to compete with other individuals for their posts to be picked up by hubs. Since mainstream journalists can obtain a summary of the thinking of the entire blogosphere by scanning a few top blogs, the integration of a story or link into a top blog may allow it to spread to the traditional media.

Thus the index authority of top bloggers is confirmed by their operating as intermediaries or brokers between the blogosphere and the political and media spheres. Journalists quote blogs. Politicians use them to communicate with activists and supporters. Barack Obama published an essay on Daily Kos, and John Kerry is a regular contributor. Further, the site's founder himself has been legitimised as a figure of recognised status. Kos is a frequent interviewee on political talk shows or news programmes, and, since November 2007, an occasional contributor to *Newsweek*. The prominence of the Kos brand was verified by the name of the annual progressive blogger conference, YearlyKos (which was renamed Netroots Nation in 2008), and by the venom which enemies pour on the name.

Originally Kos attracted an audience because he provided insightful content which had hitherto been the exclusive province of insiders. He analysed opinion polls and fundraising statistics for hundreds of congressional races around the country. This data was to be found in local newspaper polls and reports which were freely available online. The charismatic expertise of the top blogger is demonstrated by accurately predicting the outcome of political races or the impact of political events, as well as by the capacity to publicise judicious information from an obscure corner of the Internet. In doing so, Kos was revealing to the world the inside dope, the raw data which only the elite had had access to: where resources were being allocated, what the real polling numbers were, which candidate's ground game was stronger, and the like.[22] Kos added to this an aggressive style – in marked contrast to those Democrats who valued compromise over confrontation – which endeared him to his audience.

The dynamic of online charisma on Daily Kos (index authority reinforces hacker authority and vice versa) is similar to that of any successful political blog. Where the site stands out is in its blending of these inherently individual characteristics with a team of editors and an extended tribe of contributors. Kos's expertise was first reinforced by that of a dozen or so contributing editors or 'front-pagers', whose posts appear on the weblog's central text column. Front-pagers were originally authorised to publish at weekends to offer Kos relief, but he eventually gave them the opportunity to publish at any time. They include attorneys and activists as well as students. Every year a call for nominations is posted, and after recommendations from site users, a small number of people are then selected by Kos.

But more than inviting a few people to contribute to his front-page, the true impact of Daily Kos is attributable to its owner having fully integrated the 'small blog' to 'big blog' alerting mechanism mentioned above. By enabling anyone who creates an anonymous identity to post 'diaries' (blog posts) Kos wholly outsourced to the 'small blogs' themselves the job of providing his blog with interesting stories, and in so doing, tapped into a torrent of political expression.[23] Kos can thus incorporate the expertise of gifted newcomers; a common practice is to post both on one's personal or group blog and on Kos. In a sense Daily Kos truly is a microcosm of the entire progressive blogosphere and benefits not only from preferential attachment (new entrants link *to* it) but from preferential production (new entrants post *in* it). A good example in the 2008 primary was Poblano (Nate Silver), whose statistical skills enabled him to accurately predict the result of numerous primary races, turning him into an instant sensation. Poblano published comments about the election on Daily Kos and linked back to his own site, fivethirtyeight.com. Authority on Daily Kos is also measured by how long one has been a member of the community. Since every new member is attributed a user ID ('uid') number in the order of joining (for example, Kos's uid is 3, Poblano's is 110,065), participants in an online discussion or dispute can easily tell how long an opponent or ally has been present on the site.

In an environment in which new diaries are posted literally *every other minute*,[24] and in which new links to diaries edge previous ones off the 'recent diary' sidebar on the left of the front page, the question, of course, is: How does a diary get noticed, and in particular, included in the 'recommended diaries' sidebar (the 'rec list', placed above the 'recent diaries' sidebar), where popular diaries remain for *one whole day*? The answer is appropriately at the same time both meritocratic and technological: those diaries that receive the most positive recommendations are automatically included in the 'rec list'; diary authors can also give the title of 'tip jar' to a comment to encourage accolades. That getting on the rec list is a great honour is shown by the countless diarists who, when they 'make the rec list' for the first time, point this fact out in an update and thank those who recommended them. The index authority of diaries even has its historian, Jotter (uid: 3,541) who publishes daily and weekly statistics of 'high-impact diaries' which detail how many diaries were published, rated and recommended. Jotter also details the most active 'kogs' (users who write a diary, recommend a diary or comment on a diary – a more common term is 'Kossacks') and lurkers (users who only view diaries). Jotter always begins his diaries by welcoming the most recently registered new user to the site.

For a while, *Factsheet 5* served much the same purpose for zines as Daily Kos does for progressive blogs: it was the central hub on the network. The greatest prize for a publication sent to be reviewed was to be featured at the start of the magazine, in the 'editor's choice' section. This represented a big boost in visibility for a zine. Similarly, and in line with the dual nature of online charisma, diaries can be plucked from obscurity thanks to editorial intervention. Every evening 'rescue rangers' (front-pagers or invited users) present their 'pick of the day', and diaries can also be 'bumped' (promoted) to the 'front page' (the main column of text) by front-pagers or by Kos himself on a case-by-case basis if they are deemed 'interesting'. This selection of diaries by hacker authority fiat serves to bypass the structural characteristic of the blogosphere, which the Daily Kos diary field reproduces – it favours incumbents and it is hard for new entrants to break in. These mechanisms, which potentially allow anyone to rise to the top of the English-speaking world's most widely-read progressive political blog, represent an explicit challenge to traditional media's claims to professional authority. They exemplify the online information abundance which disrupts corporate dominance of information production and dissemination.[25]

This contestation of the mainstream media's status has not gone unnoticed. Traditional journalists have questioned the moral authority of blogs to speak in lieu of the mainstream. When speaking to NYU journalism students, Brian Williams of NBC *Nightly News* described his life spent 'developing credentials to cover [his] field of work, and now I'm up against a guy named Vinny in an efficiency apartment in the Bronx who hasn't left the efficiency apartment in two years'.[26] That words uttered on the screen stand on their own, irrespective of their author's identity or status, is indeed the blog credo. The fight against credentials is a familiar refrain. When Bluebeliever (uid: 54,971) wrote a diary wondering whether the front-pagers who post on defence and science issues had a background that 'makes them qualified to speak on [these issues] as an expert, rather than a casual observer', he or she got 594 responses.[27] The second commenter was the front-pager then known as Armando (uid: 1,638). He asked BlueBeliever: 'Who the fuck are you? What is your name? Where are you credentials?' – and in the ensuing comment thread, repeated the first part of the question *eight* times as well as 'who are you?', also *eight* times. When Robert in WV (uid: 57,345) accused commenters of 'newbie-bashing' the diarist, Armando replied: 'Who are you and who is the diarist. Having a posting history is the credentialism we use here. You seem not to get that. Have a zero from me for your attack.'[28] The importance of the tribe's capacity to allocate authority autonomously explains the intensity and range of proffered arguments.

Commenters wrote that the community's vetting procedures were truth-inducing; that providing great links helped to buttress arguments; that there were many distinguished historical precedents of pseudonymous writers. rm (uid: 455) invoked Aristotle's argument in the *Rhetoric* that persuasion originates in what the speaker says, not in his reputation. Armando called on the authority of Kos himself: 'Well then – You know who KOS is right? He knows who WE are. And he feels comfortable publishing us. So how the fuck is it your business then. Do you trust kos or not is the question isn't it. Here's the solution, don't read us. Leave. Now.'[29] Others derided the diarist's comment that revealing one's identity was the price for the publicity they reaped. Rick Oliver (uid: 46,966) wrote: 'there's no price for publicity, because there is no publicity. They are anonymous. That's the fucking point.' Over and over, commenters explained that credibility was earned over time, by contributing resources, and that it was not possible to import authority from elsewhere. And finally, other front-pagers joined the fray. SusanG (uid 8,411) detected behind the question a 'basic insecurity about one's own judgment', as well as an 'over-reverence for "authority", in lieu of respect for basic reasoning'. And DavidNYC (uid: 73) argued that past and present front-pagers *are* credentialed: 'We were trusted and respected enough by a community of thousands

– plus the proprietor, Markos – to be tapped for the front page. That widespread community-based trust and respect is a pretty important credential in my book.'[30] This statement sums up online charisma: the aggregated trust of a community of thousands (index authority) is confirmed by the anointment of the founder (hacker authority).

The most contentious dimension of authority on Daily Kos is the management of comments. Since hundreds of diaries are published every day, and since many of these generate dozens or even hundreds of comments, the site's administrators decided to distribute the policing of comments to the site's users. Users who have posted a large number of well-rated comments acquire 'mojo', and this index authority turns them into 'trusted users' who are allowed not only to rate others positively, but also to rate comments as '0'. A '0' rating is popularly known as a 'troll rate' (TR), because it implies the ratee is a troll. It is also called a 'hide rate', because if a comment receives three times as many TRs as positive ratings, it becomes invisible (except for trusted users). Users who receive 'too many' TRs are liable to fall prey to the 'autoban' function: they will automatically be banned from the site. One assumes that the combination of judgment by the crowd and execution by the machine is meant to indicate that the owner is in no way responsible for this most authoritarian of actions: banishment. Kos purposefully does not disclose precisely at what point one becomes a truster user, or after how many TRs one will be banned. This may constitute a protection against the system being manipulated (the stated purpose) but it also serves to shroud the processes involved in a mysterious fog, so that expulsions seem to happen almost naturally. Troll rating by trusted users to control comment quality occurs on other sites, such as MyDD. Daily Kos commenters follow threads on other sites, discuss the appropriateness of TRs and compare the TR policies of different sites, or discuss how a Kossack has commented or TRed on another site.

Community moderating represents more than emerging technologies enabling consumers to filter what they read, something Cass Sunstein argued was dangerous for democracy.[31]

Here people are going further, by actively destroying opinions they disagree with. The site's chiefs acknowledge the issue, but have decided that it is worth it: 'the efforts *here* are to define and build a progressive infrastructure, and conservatives can't help with that', writes one of the front-pagers.[32] The danger of the 'echo chamber' is real but 'a bigger danger is becoming simply a corner bar where everything is debated, nothing is decided, and the argument is considered the goal'.[33]

Usually weblogs exhibit a basic distinction between creators and commenters. As we have seen, the situation is more complex on Daily Kos, with a micro-level of trusted users who can delete comments and (indirectly) other users; a meso-level of front-pagers or 'admins' who can directly delete controversial or offensive diaries; and a macro-level at which Kos himself can ban or reinstate users. As the Daily Kos FAQ states when explaining the penalties for 'sock puppetry' (creating fake identities): 'if you are banned as a user for any reason, the only court of appeal is Markos himself'. What this means is that there is no real political *we*; no formal collective sovereign authority to appeal to if someone feels they have been wronged; no redress mechanism. Why is there no demand for more democracy from commenters? The reason is that even in situations of distributed content production and comment moderation such as those prevailing on Daily Kos, the structure of a weblog is intensely personal. Not because not everyone can be on the front page at the same time; but rather, because it is the logic of individual charisma that permeates any blog – *my* comment, *my* diary, *my* rating – and Kos can say that the whole thing is hosted on *his* server. For in the end, above all comments, above all diaries, above all front-pagers even – there is Kos. He started this, he named it, he *is* it – the daily him. The FAQ puts it rather tersely: 'First of all, no one speaks for Daily Kos other than me. Period.'[34] Ultimately in weblogs, administrative power is happily autocratic. The expertise of the crowd does not translate into the corresponding capacity to act. Micro-acts such as troll ratings, which allow the correction of another's inappropriate or inaccurate utterance, are a far cry from influencing strategic decisions such as site policy. But such micro-acts can, if occurring sufficiently consistently, lead to notable events.

Conflict: The Alegre Affair

Are weblogs useful forms of discourse, or do they run the risk, as Sunstein contends, of increasing polarisation by allowing people to engage with only those opinions that conform to their pre-existing beliefs?[35] Indeed the single most consistent finding of network analysis of the North American blogosphere is that it is strongly divided into two camps, Democratic and Republican, with little common interaction.[36] However, the polarisation argument does not sufficiently take into account a defining characteristic of political blogging: its adversarial or agonistic nature. George Packer observes that in the blogosphere 'there's a constant sense that someone (almost always the blogger) is winning and someone else is losing'.[37] But if blogging is always a criticism of someone else's opposing view, the need to refute these arguments means that the opposing argument needs to be (at least) linked to and acknowledged. Of course, a large number of such links serve to buttress 'straw man' arguments (caricatures set up the better to be torn down), but quantitative studies have found that numerous links do engage with the arguments of others, or 'at the least, politely acknowledge them as the source of some information discussed by the blogger'.[38]

Bloggers need enemies. Without them, they would have nothing to write about; that much is clear. Anthropology has shown that enemies are vital to coalesce the boundaries of exclusion and inclusion. For progressive blogs, the 'other' is divided into two camps: the traditional media and the Republicans. The traditional media flitters in and out of the blogosphere's enemy space, but apart from Fox News, blogs would really like to reform the traditional media, rather than destroy them. Under the Bush administration, the true 'others' for US progressive blogs were President Bush himself ('Chimpy'), Vice-President Cheney and their cohorts. Next came what were widely seen as their direct propaganda arms, Fox News ('Faux Noise') and reactionary talk-radio hosts such as Rush Limbaugh. Finally, the enemy blogosphere: right-wing bloggers, called 'wingnuts' or 'dittoheads' to mock their lack of autonomy from conservative talking points. Being recognised by such people as an enemy was a source of pride. After a Daily Kos diarist had interspersed photos of the wedding of one of the Bush daughters and photos of Iraqis injured or killed since the US invasion, the ultra-conservative talk-show host Bill O'Reilly declared that Daily Kos was a 'cesspool'. Using the weblog to elevate the level of hatred for the Bushes was reminiscent of the white supremacist David Duke's ideology, and as for Kos himself: the purveyor of this 'revolting website' was 'probably one of the most despicable Americans in the country'.[39] Finally O'Reilly deplored that Kos had been legitimised by Newsweek's offering him a regular column. Just as on Usenet, vigilante groups appeared on Daily Kos. Self-appointed 'troll hunters' set out to get republican trolls autobanned in order to protect 'the biggest voice on the Left'. In the process, 'civilians' got hurt – something the troll hunters regretted, but as one gruffly remarked: 'that's the harsh reality of living in the shark tank, not the kiddie pool'.[40] Troll hunters justified their actions by arguing that under Bush and his chief strategist Karl Rove, all opposition had been crushed. The 'swiftboating' of John Kerry was fresh in most people's minds.

For all their intensity, the most violent conflicts are not waged against those who are perceived as completely other. The most intense conflicts are waged against people who occupy positions spatially close to the attackers, offering a narcissistic (though insufferably antagonistic) reflection of their own image.[41] This is what happened in the North American liberal blogosphere – and most particularly on Daily Kos – during the Democratic primary in 2008. A fratricidal conflict opposed supporters of the two Democratic front-runners, Hillary Clinton and Barack Obama. The weapon in what became known as the 'primary war' was the administrative power with which users of the site were entrusted, and which allowed them to gang up on, and exterminate, one another.

Though a new generation of West Coast female bloggers such as Arianna Huffington and Jane Hamsher had made a strong impact on the blogosphere, in most observers' view, the field was still skewed against women.[42] A large number of female bloggers at the 2007 YearlyKos convention complained of being harassed online for their views on issues such as abortion.[43] On Daily Kos itself, controversies such as the Pie Fight, an advertisement deemed to sexualise women, and Kos's justification of it (he disparaged censors and the 'sanctimonious women's studies set') had shown that sexism was alive and well on the site.[44] Since then the proportion of female front-pagers had increased, though males were still twice as numerous. In general, gender issues were not resolved, but they were at least acknowledged. In contrast many black bloggers felt invisible. The elite blogosphere's apparent lack of interest in cases such as the Jena 6 or the killing of Sean Bell, in which African Americans accused the judiciary of unfair treatment, confirmed what was apparent after Hurricane Katrina: there was a lack of connection between the predominantly white blogosphere and African Americans.[45] As Pam Spaulding noted on Daily Kos, race was a 'third rail' (i.e. radioactive) topic, where 'whites are paranoid about getting their heads bitten off and blacks are tired of always bringing it up so that the issue remains mostly undiscussed with the exception of emotional eruptions'.[46] The Democratic primaries of 2008 would prove to be a catalyst for emotional eruptions.

The core of Barack Obama's brand – his inspiring slogans of hope and change, his diatribes against Washington's 'politics as usual', and his appeals to something greater than the individual, to a 'higher purpose' in political life – fit precisely the Weberian model for the charismatic leader. Yet this is not why the progressive blogs endorsed him. Obama resonated with the netroots because he had opposed the Iraq war from the start and because he was not afraid to state positions that forcefully contrasted with those of Republicans or with media conventional wisdom. The Obama campaign picked up where Howard Dean's insurgent campaign had left off not only in terms of policy, but also in terms of electoral politics. Obama's campaign appeared to embrace the self-organising, grassroots approach which Moulitsas and Armstrong had advocated in *Crashing The Gate*. Howard Dean had been elected leader of the Democratic National Committee in February 2005 and advocated rebuilding the Democratic Party base from the ground up by embracing a '50-state strategy'; the Obama campaign signalled its intention to expand the electoral map.

On these two core issues – the war and grassroots organising – Obama stood in stark contrast to Hillary Clinton. Amongst progressive bloggers, Clinton was widely perceived to embody the kind of cautious Democrat who was a member of the centrist Democratic Leadership Council (DLC) and was overly reliant on the opinion of unprincipled and risk-averse political consultants. Clinton's chief polling wizard, Mark Penn, advised her whilst remaining CEO of the Burton-Marsteller PR firm which was engaged in union-busting practices. Such insiders deemed it wisest to concentrate electoral resources on a few key swing states which would deliver the necessary Electoral College votes. Policy-wise, they embraced triangulation – the art of adopting Republican positions in order to insulate oneself from Republican attacks, a tactic utilised to great effect by Bill Clinton in the 1990s. Triangulation's blurring of differences between the parties led to its unequivocal condemnation by progressive bloggers, who pointed to the support for the Iraq war by DLC 'chickenhawks' as a prime example of such 'inside the Beltway' foolishness.

Hillary Clinton had never gathered more than 10 per cent of support in the polls conducted on Daily Kos. This dislike appeared to be reciprocal, as evinced by leaked remarks in which Clinton disparaged Democratic activists and in particular the MoveOn organisation.[47] For their part, many Clinton supporters ('Clintonistas') were sceptical of Obama's rhetorical flourishes; they viewed him as untested, an empty suit into which everyone was free to project their own hopes and interests, in contrast to their preferred candidate's proven resistance to right-wing attacks. The divide was generational, between so-called 'baby boomers' who tended to favour Clinton and the so-called 'generation Y' (18–35 year olds) supporting Obama.[48] Clinton also represented the first real chance a woman had ever had of reaching the ultimate political prize. The adulation expressed by audiences during Obama's large rallies inspired Clinton supporters to derisively label his supporters (clearly members of a 'cult') Obamabots, Obamadroids, Obamazoids, the ObamaBorg – the latter a reference to the villainous *Star Trek* all-in-one collective entity. The primary war was fought in countless diaries. What follows is a snapshot of a few significant battles. The first skirmishes occurred in the lead-up to the Iowa primary. Alegre (uid: 47,824) was a typical Clinton supporter: she was female, in her forties. What distinguished Alegre were her energy and enthusiasm (she posted a diary every day), and her willingness to use weapons such as sarcasm and profanity. She thus became a champion for Clinton supporters and a lightning rod for opponents. In mid-December 2007, when she queried in her diary why 'flaming turd throwers' turned up on a nightly basis, Geekesque (uid: 59,129) replied, 'Maybe if you didn't post such flaming turds of diaries you wouldn't get that reaction'. Geekesque then taunted: 'Contact an admin and ask to have me banned. I think if they refuse, you should post a diary threatening to leave.'[49] Tensions continued to escalate through December, and eventually Kos had to step in: 'We know for a fact that hardcore partisans are coordinating to uprate troll comments and down-rate non-troll comments according to their viewpoints', he posted on 27 December. This would not be tolerated; people engaging in such activity would have their ratings ability 'yanked' and these powers may or may not be reinstated after the primary.[50] On 28 December a Hillary 'roll-call diary' was posted, allowing Clinton supporters to network and count themselves. In the previous days similar Obama and Edwards diaries had included requests in the comments to refrain from attacking the candidates, and a warning by the diarist that any such negative discourse would result in a troll rate; this had gone unchallenged. However when similar requests and threats were made in the Clinton diary, flubber (uid: 54,774) commented: 'Make my day – Troll rate away [...] Perhaps I've missed the section of the site rules that lets you declare martial law in your diaries. If so, you'll have to enlighten me.' Front-pager Kagro X then informed the diarist that he or she had no right to use threats to pre-empt criticism. This was strongly contested by another female Clinton supporter, Goldberry (uid: 42,723):

Clintonistas are frequently at the receiving end of a TR and our diaries are at the mercy of many marauding Edwards and Obama people who are just itchin' for a fight. I've had my frickin' **Tip Jar** TR'd. It's a relief to not have to fend off disrupters whose sole purpose seems to be tormenting us and driving us out of our own tribe. If Obama wins, do you want our help or not? So, if you don't mind, go haunt some other diary for awhile, Kagro X. I think power has made you giddy and your partiality is showing. Don't you have a front page diary to write or something? I think I hear your momma calling.[51]

Eventually Kos intervened to reiterate the point: any kind of pre-emptive censoring was unacceptable. Since Kos's appearance in comment threads are not that common, they carry a certain weight, and this effectively sealed the argument. When asked why it was non-problematic for the Obama and Edwards diaries to feature this warning, but not for the Clinton one, Kagro X replied that he had not seen the previous diaries, and retrospectively condemned them. Yet such perceived unfair treatment contributed to sowing the seeds of division.[52]

Whatever substantive policy and political differences might have existed between the candidates' camps soon became intermingled and in some cases submerged by identity politics. Mutual accusations of sexism and racism rent the site apart. There can be no doubt that Senator Clinton's campaign encountered terrible sexism: hecklers shouting 'iron my shirt' at rallies, a focus by the mainstream press on the candidate's cleavage, on the way her laugh resembled a 'cackle', were examples of the way in which the candidate was sexualised and belittled.[53] Obama supporters argued that none of this was their candidate's fault. True, they continued, he did not denounce sexist statements made in the media: surely it was enough to disassociate himself from people in his campaign who had made offensive remarks? No. As Michelle Goldberg put it, for feminists seeing Clinton lose to a younger, more charismatic man seemed to echo a primal experience of middle-aged female humiliation.[54] In any case Obama supporters countered with accusations of racism, pointing for example to

Clinton supporter Geraldine Ferraro's assertions that Obama owed his success to his race.

White privilege, male privilege. Each faction accused the other of using surrogates or coded messages to convey racist or sexist messages meant to disparage their candidate or imply that he or she had an 'electability' problem. When opponents responded that they were being oversensitive, complainants would emphasise that people on the other side were incapable of understanding offensiveness which did not concern them; were, in fact, blind to it, 'we get it that you don't get it'. Each camp accused the other of behaving like Republicans and of stoking the fires of divisive identity politics. Each accused the other of taking for granted the constituencies which had historically formed the base of the Democratic party. Both charged that their candidate was being held to unreasonable standards for being declared the winner because of their ethnicity or gender.

On Daily Kos, it wasn't so much sexism, more undiluted vitriol. In a post beseeching Kossacks not to troll rate comments they disagreed with, RenaRF (uid: 37,061) gave the following counter-examples of comments which – in her view – had been justifiably troll-rated, and that she had subsequently pulled out of the 'hidden comments'. They offer a taste of what Clinton supporters had to put up with: 'Let's get the Hillary turd flushed down the toilet so we can get down to the business of destroying McCain.' [snip] 'Hillary Clinton is a god damn bitch. This sorry excuse for a human being has now destroyed the democratic party. Hillary you can go straight to hell!' [snip] 'Please go fuck yourself.'[55] When Clinton announced that she would release her tax returns, slinkerwink (uid: 4,335) posted a call to create teams to pore over them.[56] As reported by Todd Beeton at MyDD, many self-confessed Obama supporters voiced their opposition to such tactics.[57]

On 8 March 2008, Alegre posted an article celebrating International Women's Day in which she explained that she couldn't think of a better way to celebrate the occasion than by 'recognizing one of the most well known and respected women of our time, Senator Hillary Clinton'. The diary attracted positive comments about the senator's accomplishments, but also criticisms of her vote for the Iraq Resolution of 2002, amongst other topics. When another commenter complimented Alegre on a well-written and informative diary, she replied: 'There was a time when one of these diaries would have raced up on to the Rec List. Sadly, those days are gone here. People have left and the place is dominated by bullies and haters.'[58] It has to be said that Alegre did not always help her own cause. On 11 March she posted a diary dealing with Geraldine Ferraro's statement that Obama was very lucky to be black. Several commenters pointed out that a week earlier Alegre had strongly complained about an aide to Obama calling Clinton a 'monster', whilst now she described Ferraro's statement as 'not worded right' and urged everyone to 'move past this and start talking about oh I don't know ... the issues for a change?' She was duly described as a 'shill', a 'hypocrite' and as 'defending the indefensible'. By 14 March, Alegre had had enough. She announced to the world that she was leaving Daily Kos:

> This is a strike – a walkout over unfair writing conditions at DailyKos. It does not mean that if conditions get better I won't 'work' at DailyKos again. Because of administrative inaction our community has become little more than an echo chamber with an attitude that harkens back to the early days of Dubbya's administration – *yer either with us or yer a'gin us, heh!* ... The double standards, the distortions, the hateful, irrational, personal attacks, and the lies about Hillary and her long and distinguished record of public service stop here – and they stop now.[59]

Asserting that this was a 'laughable strike', on 17 March Kos quoted a post by respected analyst Al Giordano which declared that well-cultivated blogs have to be 'weeded from time to time'. Deciding that it was time to clarify where he stood, Kos summarised the differences between the candidates. While acknowledging that they probably both relied on consultants, he went through the familiar issues (the war, the 50-state strategy, the DLC) , all of which were dwarfed by Clinton's cardinal sin – *losing*. His own declaration of war followed: it was Clinton, with no reasonable chance of victory, who was fomenting civil war so as to overturn the will of Democratic voters. Therefore, she did not deserve 'fairness' on Daily Kos. Sexist attacks would not be tolerated, but otherwise 'Clinton has set an inevitably divisive course and must be dealt with appropriately'.[60]

Alegre did not leave alone – 68 other Daily Kos users added their names to her declaration. Their voices would be heard over the next few months, in the comment threads of further Alegre diaries. Others went too. Many had already left. Where did they all go? The fork did not constitute one unified project; it splintered across the progressive blogosphere, sending shrapnel of hatred back towards Daily Kos at every occasion. Alegre and Gabriele Droz created *hillarysbloggers*, from whence they would cross-post to other sites; and in June 2008 *Alegre's Corner* was set up. They were following in the footsteps of Goldberry, another vocal Clinton proponent, whose last diary on Daily Kos (which made fun of the religious zeal of the Obama movement) had generated mutual accusations of racism and sexism.[61] Under the name Riverdaughter she launched a new blog for those 'pushed or voted off the Daily Kos island'.[62] Refugees from Kos who chanced upon it could then exclaim in the comments: 'AHA! I've found you all. What a happy day. Great blog and much love to everyone. Dkos has become a vile stinkpot. I feel like I've stepped into heaven.'[63]

The virtual diaspora also found refuge on pre-existing pro-Clinton blogs such as TalkLeft and MyDD. After James Woolcot published an article on the rift in *Vanity Fair*, TalkLeft commenters reminisced about their experiences.[64] They described troll-rating abuse so severe that it resulted in a party purge, in excommunication, in hate. FlaDemFlem explained why she had had no choice but to leave:

> I left DKos because I got tired of being called names like bitch, slut, moron, stupid, idiot, etc. I got tired of being accused of being a paid Clinton poster. I do not get paid for posting to blogs. If I did, I would post to more than one or two, I can assure you.[65]

Commenters at *The Confluence* also described their treatment as abusive and misogynistic, writing that they could not take it any more and left despite 'having a uid in the 16,000s' (Eleanor

A) or '18,000s' (1040SU), whilst others did not attribute their leaving to sexism but to their support of Clinton, resulting in their being labelled 'corporatists' or 'right-wing Trolls' (litigatormom, Eleanor A).[66] The lack of administrative control was contrasted to the positive influence of moderators on TalkLeft: 'What I like is that there are no gangs of TRers. My god the censors were mad over there' (salo). The double standards when dealing with different groups was also invoked: 'When Obama says nice things about Republicans, he is riding the Unity Pony. When Hillary says anything at all that someone characterizes as Republican, she is ridden out on a rail' (litigatormom).[67] Commenters were suspicious that many leading pro-Obama bloggers, such as Kos, John Aravosis of Americablog and Arianna Huffington, were former Republicans. Were they trying to 'hijack' the party for nefarious purposes? What right had they to define who was and was not part of the tribe? Commenters on TalkLeft were critical of Kos's motives: 'He has become much too big for his britches and drunk with power, not unlike Rush Limbaugh ... I believe it's because KOS fancies himself to be a king maker. His ego was getting out of hand before the primary season ... The tendency at DKos is movement toward a totalitarianism that would make Orwell shudder ... conformity to the leader, Kos or Obama, is what defines what is right.'[68] The Great Orange Satan, a term originally used by conservatives because the Daily Kos banner and hypertext links are orange, was now variously known as Orangebama, the Big Orange Sippy Cup, the Big Dark Orange Place, the Big Orange Frat House.

And what of Alegre? She practised her craft in a variety of sites, but most prominently on Daily Kos's twin site. MyDD's founder, Kos's erstwhile co-author Jerome Armstrong, had come out as a Hillary supporter because, as he explained, 'Clinton's got a closer resonance with what progressives need now as a President'.[69] Yet even in such a favourable environment, Alegre eventually ran into opposition. The tide seemed to be turning on MyDD and her cause was not helped by the last throes of the Clinton campaign. It was hard to spin Hillary Clinton's remark in May 2008 that she was supported by 'hard-working Americans, white Americans' in a positive way, for example. A joker sarcastically named Hillary-willwin remarked in early May: 'Alegre with the Obama people taking over Mydd and being mean to Clinton people like dailykos was don't you think it is time to strike at Mydd?'[70] In fact, Alegre felt compelled to address herself directly to the site's administrators: once again, double standards were being used. Hillary was fair game but any criticism of Obama's electability problems (his controversial pastor Wright, his unsavoury patron Rezko, his radical friend Ayers) meant Alegre and her friends were tagged as 'disloyal to our Party, turncoats, GOP Operatives, Karl Rove mouth-pieces, or worse yet – racists. So tell us what is allowed?'[71] Eventually chrisblask offered the following response:

> What IS Allowed: Discussion. Not even 'allowed' – that carries an authoritarian tone that only Jerome can answer to – but perhaps 'commonly desired'. There has certainly been nothing stopping you from posting and dominating the reclist, so obviously you are allowed to say anything you like. What I think is commonly desired is discussion. If you were to engage in the comments section of your diaries, then both those who agree with you and those who don't could engage you in the art of social cognition known as debate.[72]

This effectively ended the thread (which featured 470 comments). What Alegre was being charged with, was, in essence, breaking the rules of communicational rationality: where was the sincere effort to empathise with an opponent's views? Alegre had indeed perfected the art of the snappy rebuttal of *certain points*. More than anything, it was this selective responsiveness that infuriated her opponents. A litany of posters on MyDD had politely asked: Are you going to answer my question from four days ago – and mine – not to forget *mine*? In a weird echo of the scenes which had echoed down the Great Orange Corridors, they proceeded to pour scorn on her: Alegre, they said, was intellectually dishonest. She never addressed matters of substance. She selectively edited stories to suit her bias – she was probably a paid Clinton operative. And she got her buddies from Hillaryis44 (a virulently anti-Obama site) to swamp comments to put her on the rec list.[73] Yet they could not stop rising to her bait; she still spoke for so many! Like

Hillary, she played by her own rules. Like Hillary, her existence was an affront to some, a beacon for others. After Obama secured the nomination on 3 June 2008, ringing calls for Democratic unity issued. Diaries urging Alegre to 'come home' were posted on Daily Kos. Alegre? She was not interested. Like the Japanese soldiers who kept on fighting in the jungle long after their country had capitulated, she refused to give in, still vowing that Hillary would take the fight to the Democratic Convention in Denver, and when that did not happen: to 2012. She had become a symbol, standing for something larger than herself that would never, ever give in – consequences be damned.

7

THE IMPERFECT COMMITTEE: debian.org

So that I can continue to use computers without dishonor, I have decided to put together a sufficient body of free software so that I will be able to get along without any software that is not free.

Richard Stallman, *The GNU Manifesto*

In March 2004 a female developer raised a question during the annual debate on the debian-vote email list, where candidates for the position of Project Leader field queries: How could Debian be the 'universal operating system' when the project hardly featured any women; were there any plans to redress this gender imbalance?[1] Another woman added that she often felt apprehensive or intimidated because of the expectation that men would note her gender and automatically assume her to not be as good at computing. In the ensuing debian-vote thread, though some developers were generally supportive, others called her a 'flake', 'mentally unstable' and 'sexist'. A developer also posted a snippet taken from an Internet Relay Chat (IRC) channel in which a woman developer was subjected to puerile banter.[2] A few months later, the debian-women discussion list was established, with the aim of increasing the role and visibility of female developers within the project.

The premise of Free and Open Source Software (FOSS) development is that intellectual property rights should not hinder the creative ability of computer engineers ('hackers') to freely improve whatever piece of code they want to play around with. Further, with licences such as the GNU General Public Licence, contributors know that their work cannot be absorbed into proprietary software. Alongside technical excellence, an important but less recognised dimension of hacker-charismatic authority has to do with the defence of one's honour. Debian is overwhelmingly male, and anthropology has shown that honour is the foundation of the patriarchal order.[3] What is the impact of honour on Debian? Does it clash with the tribe's stated norms of behaviour? Might it thus compromise what is the aim of all FOSS projects, the construction of a more perfect system? This normally concerns only computer code, but in the case of Debian, developers have also sought to build a complete *political* system.

Project: The Universal Operating System

Debian was initially announced on 16 August 1993, when Ian Murdock posted his intention of creating a Linux distribution (a complete operating system and series of applications) to the comp.os.linux.development Usenet newsgroup. The aim was to develop a 'commercial grade' but non-commercial version of the GNU/Linux operating system (OS) which would be easy to install and contain the most up-to-date versions of 'everything'. Instead of focusing solely on the basics – a kernel (the OS core), utilities and development tools – Debian would be intended for a bigger audience than developers and would feature a window system, document formatting tools and games. The project would also include extensive documentation, something hackers are not usually very interested in working on.

In line with the canonical view in FOSS circles that 'really great hacks come from harnessing the attention and brainpower of entire communities',[4] Murdock reflected that Debian's most important contribution to the world was its decision to adopt a community-based development model: 'As far as I know, this marks the first time that a project *intentionally* set out to be developed by the community that used it … After all, if you remove the community from open source software, it's just software.'[5] As with many FOSS projects, participants strongly identify with the common cause. Their feeling of belonging derives from shared social and political beliefs, such as the value of free access to information and open communication, and shared esoteric knowledge, such as the C computer language and the Linux variant of the Unix system.

Their collective identity is also made up of 'values, traditions and an endogenous written history'.[6] This international community's remarkable achievement in building such a complex system was made possible by the Internet's effectiveness as a distribution platform. Debian's core mission is to be 'the universal operating system'. This signifies that it can be used 'for most anything, on most any hardware, and [can] install most any software'.[7] A distinguishing feature of the Debian distribution is that it comprises over 18,000 binary packages available for instant installation. There is something encyclopedic about this desire to do everything. But size matters less than quality. And here one can say that for the true believers Debian is the Mary Poppins of operating systems: it is *practically perfect in every way*.

Proponents would first argue that Debian is *technically* superior to proprietary software. No surprises there: this is free-software orthodoxy. Companies such as Microsoft could not possibly afford the number of testers and developers necessary to create flexible, secure, reliable, cheap and innovative software 'appropriate to the vast array of conditions under which increasingly ubiquitous computers operate'.[8] Hacker folklore has it that Microsoft itself recognised the superiority of FOSS. In the famous 'Halloween memo', a Microsoft hacker opined that Linux's open-source code gives it a long-term credibility which 'exceeds many other competitive OS's'.[9] According to the Debian project's secretary, the Debian star has a special brilliance within the FOSS firmament: in terms of *community*, unlike other Linux systems, Debian has no 'caste system' of core developers looking down their noses at 'lowly newbie wannabe contributors'.[10] Further, Debian has the best *translations* into most languages; and its *bug-tracking system* is better than those of all other UNIX systems. Then there is software *maintenance*, still an important component of any system administrator's job. With Debian? 'It's simply trivial. It's a nonissue. Don't even bring it up when talking about any problems with Debian, it's not worth the effort. Borderline flawless.'[11] Once again: practically perfect! The key to Debian's robustness is its modularity, following the Unix philosophy. More than 1,000 developers work on different aspects at the same time. Each one maintains his own 'package' or modular component thanks to the package system (dpkg), which allows the system to be upgraded piece by piece. Strict guidelines ('policy') allow the packages to be independent but inter-cooperating. The result, for the untrained eye, is a bewildering array of packages.

This derives from Debian's strict adherence to the *philosophical* principles of free software, autonomy and transparency. Developers and users can configure Debian exactly the way they want it: 'you control the system, and not the other way around'.[12] Debian also has the best *legal* licence, the Debian Free Software Guidelines (DFSG), written by Bruce Perens and later adapted to become the official open source licence. Debian even decided to reject the Free Software Foundation's GNU Free Documentation License, because it contained 'invariant' sections which cannot be remixed, thus contradicting the spirit of free software.[13] Other FOSS projects have their own licences, but until Debian came about few had a *constitution* or a *social contract*, to which all participants must adhere. The social contract states that Debian will remain 100 per cent free (according to the DFSG); that it will give something back to the free-software community; that problems will not be hidden; and that the project's priorities are its users and free software.

Autonomy is enshrined in the constitution, which states: 'A person who does not want to do a task which has been delegated or assigned to them does not need to do it.'[14] Debian thus has the best *political* system, which attempts to balance democratic sovereign procedures and charismatic meritocratic hacker skill. Debian's chief (the project leader or DPL) is elected every year. All developers can stand for election by posting their platform on the Debian website and participating in a series of debates on the debian-vote list, where they field questions from other developers, as well as in debates on IRC channels with the other candidates. Leaders are elected by sophisticated voting procedures based on the Condorcet method, in which all options are subject to pairwise comparisons to all others; the option which is systematically preferred is the winner.[15] Voting procedures are intended to protect minorities, and thereby to prevent forks: specified majorities are required for many decisions. For example, a supermajority of 3:1 is necessary for the supersession of foundational documents, such as the Debian Social Contract or the Debian Free Software Guidelines. The Debian voting system requires that continuing a discussion must always be a ballot option; statistically, 70 per cent of the resolutions require a supermajority.[16] Debian's authority structure is examined more fully in the next section.

Finally, Debian has strived to maintain its high standards when it comes to *recruitment*. The project's growing success gave rise to the need to protect quality. But inclusion filters, meant to slow the flow of new entrants, had to be managed delicately so as not to discourage applications. The movement of new entrants from the periphery to the core follows an elaborate initiation process aiming to ascertain their ideological conformity to the project (new maintainers must adhere to the Debian social contract), their social connectedness (they need to be sponsored by developers), and their technical ability (through previous contributions such as managing a package, writing documentation pages, or debugging, testing and patching). The sense that only the best need apply has a self-reinforcing quality. This comes at a price: though the Debian community is often described as unparalleled in terms of its dynamics and competence, it is often 'badmouthed as arrogant and too idealistic for the real world'.[17]

Authority: A Bazaar of Cathedrals

'The Cathedral and the Bazaar' is the title of a famous essay by Eric Raymond. Raymond contrasted the secretive low-frequency release model of Richard Stallman's Free Software Foundation's GNU software to the 'release early, release often' model pioneered by Linus Torvalds with Linux. Nicolas Auray has argued that Debian represents an attempt to make the bazaar *viable*, with norms aiming to reduce tensions, but also *moral*, with institutions aiming to reduce unequal relations.[18] This represents an evolution for FOSS projects from purely charismatic models such as Linux which are based on a 'lazy consensus': if no one makes an objection after three days, then a proposal is accepted.[19]

Debian must reconcile the central notion of each developer's autonomy, and the respect for difference, with the constraints deriving from the production of a complex system with quality standards of the highest order. This is the main purpose of the modular structure: to give developers full administrative control over their packages or teams, in a mini-cathedral model. This level of personal control is in fact a primary attraction for many FOSS developers, who are then free to work alone if they wish. There is also less chance of being criticised for one's work if one keeps control over it and only releases finished versions, thereby stopping others from interfering. At the same time, Debian packages all follow strict production guidelines ('policy') and can easily become integrated.

The absence of monetary rewards in Debian guarantees that reputation benefits earned by technical excellence (hacker charisma) are the standard for all value. One sign of authority in Debian might therefore be the number of packages any one developer is responsible for: the higher the number, the greater the hacker-charismatic authority? Yet a higher level of administrative authority is required to deal with infrastructure that stretches over different packages. Martin Krafft observes that these core tasks are undertaken 'by a smallish number of developers that take Debian very seriously'.[20] Studies of other free-software projects have shown that small groups are responsible for the majority of work. Lerner and Tirole argue that a tiny minority make the largest contributions, their integration into the 'core group' of developers representing the ultimate recognition by their peers.[21] A study of the development of Apache showed that out of 400 developers, the top 15 contributed between 83 and 91 per cent of changes, whilst 'bug' (or problem) reports were much more evenly distributed.[22]

Martin Krafft notes that since the membership of the security team is seen as prestigious,, 'some people write lengthy emails explaining why they should be picked'.[23] Naturally these requests are never honoured: only those who have contributed are deemed worthy of inclusion. Highly specialised knowledge of the project's infrastructure or of the workings of a core team, accumulated over years, risks becoming fossilised. This is the central question for Debian, and indeed for any volunteer project requiring high levels of expertise: how do the tribal elders transmit their wisdom? As will be shown in the next section, this question is largely unresolved.

Supporting new users is a crucial activity if the project is not to wither away. Hacker-charismatic authority can be detected by examining interactions in lists, the project's primary communication and education tool. FOSS lists are not simply forums for discussion, but are also means for peers to evaluate the quality of code or advice. A question must be useful for everyone, otherwise it risks the indignity of the questioner being directed to documentation such as the FAQ.[24] Studies of patterns of questions and responses on the debian-french user list reveal a tension between a system of massively distributed collective authority in which everyone and anyone authorises themselves to respond to a request on the one hand, and the constitution of an elite based on reciprocal approbation on the other. The selection of *who* one responds to, as well as *how* one responds, is crucial: authoritative responses on threads are primarily addressed to previous respondents, rather than to the original questioner. The threaded nature of discussions on lists enables the selection of high-status partners. A mutually responding core emerges, which preserves expert authority in the midst of a dynamic and open system.[25] On developer lists, studies show that the necessary knowledge of all the elements of the distribution, allowing people to foretell problems and remember solutions, is not equally distributed. List archives can be viewed as the minutes of a permanent assembly, which is directly accessible to all at every minute, but only the most experienced members can remember or find their way around the mounds of archived information. The authority of such tribal elders represents the means of moderating the obsessional reference to technical excellence.[26]

Unlike on Daily Kos, where the two variants of online charisma mutually reinforce one another, hacker and index charisma on Debian are fundamentally antinomic. This is because index authority always contains an arbitrary element: the identity of earliest entrants is due to chance, not talent. Since Debian is based on meritocratic skill, it should in theory have no place for the other form of online charisma. How did the index-authority virus enter the project? As previously mentioned, a key goal for leaders of online tribal projects is to distribute administrative power whilst maintaining quality. Since the autonomous structure makes it very hard to discipline people, the maximum effort must be borne upstream, before recruitment occurs. In practical terms, this boils down to: who can be trusted to access the levers of control? Contrary to weblogs (where administrative power is never fully opened to everyone) and to wikis (where all modifications are instantly reversible, and do not affect the whole project), contributors to Debian really do have the potential to harm the software.

As a result, a central difference between Debian and the other cooperative projects examined in this book is that contributions cannot be anonymous. Online communities are routinely described as having fluid boundaries and shifting members and identities. In contrast, members of FOSS projects who are developing software that is hosted on protected servers connected to the Internet must maintain a distinct and trusted identity, which will enable them to gain access to these protected resources.[27] Since developers are capable of modifying the code, the project has to be protected from Trojans (hidden bugs or viruses) as well as from well-intentioned but unskilled developers.

There are various means by which a digital identity can be authoritatively linked to the entity it claims to represent. The answer for Debian lay in the use of keys (large numbers) that allow data to be encoded and decoded. If the key is secret, and the same key is used to encode and decode a message, sender and recipient cannot exchange a secret key to begin with. A possibility is public key infrastructure, where a centralised certifying authority registers users and delivers certification to them. But autonomous projects require a distributed solution. The aim of securing identities led to the establishment of a private-key cryptography system, whereby a public key encodes data, and a completely different key decodes the data. The authenticity of the communication's content is guaranteed, but this process does not verify the link between the key and the sender's identity. Real-world identities must be connected to a given public key. This is achieved when identity certificates (including public keys and owner information) are signed by other users who are themselves known and trusted by the tribe. The physical act of vouching for the link between a public key and the person or entity listed in the certificate takes place at offline 'key-signing parties'. Developers bring a copy of their public key and valid photo identification; they meet and certify another's public key. A key which has been signed can then be placed on a central key server maintained by a keyring coordinator. In social network analysis terms, O'Mahony and Ferraro write that the resulting web of trust rests on the assumption that 'the more people who have signed each other's key (the greater is the density of the network), the more reliable is the information authenticated'.[28]

This process contains the hallmarks of a familiar scenario in which entrants who have accumulated many endorsements are advantaged and it is hard for new entrants to break in. O'Mahony and Ferraro have analysed the dating of key-signings and suggest that between 1997 and 2001 the Debian keyring network increasingly conformed to a power-law mode and became centralised. Formalised membership processes, such as a vetting team (the New Maintainer Committee or NMC) and the requirement of sponsorship by an existing developer, encouraged preferential attachment to gatekeepers, as measured by key-signings. O'Mahony and Ferraro's conclusion, that becoming a central player through physical participation in key-signings 'enhanced the probability of attaining a gatekeeper position far more than the number of packages maintained',[29] seems unassailable. However their assertion that preferential attachment influences the 'structure of the network as well as the design of governance mechanisms' assimilates recruitment procedures to overall project governance. There is scant evidence that this is the case, beyond the fact that the head of the NMC was elected Debian project leader in 2003. It is arguable, for example, that membership of Debian's Technical Committee (TC), or of key infrastructure teams, is of greater significance to the project, because these positions are not renewed every year.

The influence of index-charisma is counterbalanced by Debian's embrace of *sovereign* authority. An egalitarian or collectivist organisation based on the sovereign will of all participants is the natural format of 'successful anarchist communities'.[30] This takes a number of institutional forms. The autonomous authority of individual developers over their packages is overridden by the greater good, as expressed in two institutional mechanisms: the General Resolution Protocol and the Technical Committee. Though Debian *users* can take part in mailing-list discussions, only *developers* can vote in a general resolution or sit on the TC. Debian Developers (DDs) elect the Debian project leader every year. The DPL in turn appoints or reappoints the chairman of the Technical Committee and the secretary; these three roles must be distinct. The DPL also appoints delegates, such as the FTP master, who controls uploads of packages; release managers, who are in charge of supervising the release process; security team managers, who coordinate security issues with other projects; and Debian account managers or DAMs, a very sensitive position as they maintain Debian accounts and oversee the recruitment of new members, and can also expel or suspend developers. Often delegates are reappointed in their roles for many years.

The TC is a kind of supreme court which is supposed to arbitrate matters of technical policy, decide where developers' jurisdictions overlap, and, when necessary, overrule developers. The secretary administers, and reports on, the voting process. Secretaries can stand in for the leader (as can the TC chairman) and they adjudicate disputes about interpretations of the constitution. The ultimate authority is the democratic will of all the developers, who can recall leaders, reverse decisions by leaders or delegates, or amend the constitution through a general resolution. In practice however, this is seldom used: in the ten years following the adoption of the Constitution in 1998, there have only been twelve general resolutions.

As for DPLs, their authority is qualified. The constitution states that, just like a traditional tribal chief, 'the Project Leader should attempt to make decisions which are consistent with the consensus of the opinions of the Developers'.[31] Leaders make decisions for which no one else has responsibility, but should 'avoid over-emphasising their own point of view when making decisions in their capacity as leader'.[32] How can project leaders enforce their decisions? Their compliance arsenal is limited, as demonstrated when the DPL, in an effort to compel a recalcitrant developer to report on what he was up to, published an open letter to the developer in an attempt to shame him into action.[33] Because of the delays in processing new applicants, the status of Debian maintainer, situated between developers and users, was created in 2007. This authority level would be for users who had been maintaining a package for some time; they would be allowed to upload without going through maintainers. However, they would not have voting rights, nor would they have access to the debian-private mailing list or the Debian infrastructure.[34]

Conflict: the SL Saga

As has frequently been noted, external enemies serve to coalesce boundaries of exclusion and inclusion. The originating enemies of all free-software projects, which motivate and sustain their existence, are entities which restrict hacker autonomy. Lehmann writes that such enemies include 'anything and anyone which is perceived as prohibiting access, including copyrights, patents, and secret source codes, but also mechanisms that encourage dependence'.[35] Richard Stallman once declared that persecuting the unauthorised redistribution of knowledge by robot guards, harsh punishments, legal responsibility of ISPs and propaganda is 'reminiscent of Soviet totalitarianism, when the unauthorised copying and redistribution of samizdat was prohibited'.[36] This kind of outburst is rare. Hackers in general, and Debian developers in particular, do not as a rule disparage proprietary software; its inferiority is taken for granted, not dwelt upon. This is because, as in the blogosphere, conflicts with the 'same' far outweigh in intensity those against the 'other'.

A prime candidate for the role is a rival software distribution named Ubuntu. Financed by Mark Shuttleworth, a dot-boom entrepreneur, and his firm Canonical, Ubuntu's strongpoint is the clockwork regularity of its releases: every six months, Ubuntu developers 'freeze' Debian, make a selection amongst the myriad Debian packages, and release them as an integrated system. Ubuntu's ease of installation and its successful generation of a community network of support and development, complete with mailing lists, have proved popular. Debian's original founder, Ian Murdock, who left the project, commented that Debian had 'brought this fork on itself with its glacial pace'.[37] The Ubuntu website states (emphasis added) that 'the Ubuntu project *attempts to work* with Debian to address the issues that keep many users from using Debian'.[38] A developer justified leaving Debian for Ubuntu because he was tired of the frequent flame wars and because 'having one person who can make arbitrary decisions and whose word is effectively law probably helps in many cases'.[39]

Conversely, some Debian users affirmed their distinction as true free-software aficionados: 'most people who use Ubuntu (Not to insult them) are teenagers who want to use Linux in the same way "hip" people use Macs'.[40] Sometimes the hostility moved offline, for example at the annual Debian development conference debconf6 where someone 'was attacked for wearing an Ubuntu T-shirt … while someone else was applauded for wearing a "Fuck Ubuntu" t-shirt'.[41] Ubuntu was everything Debian was not: it was timely and easy to install. In addition, it was disturbing for volunteers to see others being paid for what they themselves were doing for nothing.[42] The rise of Ubuntu also risked turning Debian into a supermarket of components with little work being done on the crucial elements that work across packages.[43] But the biggest problem was the perception that as Ubuntu grew, it was effectively leeching off Debian, but not paying Debian back in the coin of the realm: improvements to the software. The Ubuntu website declares that bugs listed on the Debian Bug Tracking System (BTS) and fixed in Ubuntu are automatically communicated back to the BTS. Yet in 2008 a former Debian project leader declared that developers were unhappy about the relationship with Canonical, who they believed was not actively contributing back to Debian: 'They're not giving back as much as they claim to do.'[44] In breaking the development code, Ubuntu was not behaving honourably.

What is the role of honour in Debian? Nicolas Auray argues that the ethic of the Debian project is a form of 'moral heroism' or 'civic totalitarianism' required by the participants' constant mobilisation via the mailing lists. This generated a specific Debian 'netiquette'. Members of the tribe have to demonstrate perfect self-control, temper their emotions and follow norms of humility.[45] That inventors require norms of humility to moderate an intense focus on originality and priority had been posited by Robert Merton in his sociology of the scientific field.[46] In Debian, humility serves to temper not priority, but authority over packages. Yet it is debatable whether humility really influences behaviour in an environment predicated on technical excellence, and which clearly constitutes a continuation of the confrontations of Usenet, as evidenced by countless references to 'flame wars' and 'killfiles' (and their distinctive *plonk* noise). In this context *the threat of dishonour allows the lack of observance of the humility norm.* Honour is less a bug in the system than a fallback mechanism, an (archaic) justification for autonomous conflictuality. When a French user dared to criticise the lack of responsiveness of developers to user needs, a developer replied: 'What am I, your servant? We are volunteers who have better things to do than listen to the inept ramblings of a minority of users who know better than us what they want to do.' And the next day another developer added, speaking to the same user: 'Contrarily to you, we don't just watch, we play'.[47]

The developer's honour is rooted in being active and autonomous. An exemplary example of the role of honour was the SL affair. Several points remain unclear to the outside observer, as some of the events under consideration unfolded via IRC, in restricted-access lists, in private emails, or in face-to-face meetings at developer conferences. Nonetheless, what can be garnered from the public discussion lists is revealing of the tensions in Debian's authority structure. SL was a developer who entered into a series of confrontations with other developers; as a consequence, some of his privileges were removed. His outraged conviction that a grave injustice had been committed led him to escalate his appeals to the entire project.

On 9 March 2006 a violent exchange with another developer about the presence of a bug in a package resulted in SL urging the other developer to admit that he had been 'wrong'. A few hours later SL added: 'You prefer to keep your nonsense around, and obtusely continue to claim there is a kernel bug when it has been proven it is not the case.' This message ended with SL charging the other developer with being 'clueless' and that his words and actions demonstrated that 'not only my judgement, but also that of most other debian developers is far superior to yours'.[48] When reminded by former project leader and prominent Technical Committee member IJ that he had been asked to keep his comments 'civil and technical', and had failed to do so, SL responded that his opponent had made no technical points, instead using FUD (Fear, Uncertainty and Doubt), and concluded this time with 'I don't understand how [he] could have become a DD, and fooled the task-and-skill test'.[49] Repeatedly casting aspersions on the expertise of a fellow developer and using the acronym FUD (a term usually associated with Microsoft) is violent behaviour in a FOSS environment. This exchange motivated IJ to write a message that same day to the tech-cttee email list, titled 'Flamewars and uncooperative disputants, and how to deal with them': his opinion being that in disputed cases the attitude of the complainant or maintainer (whether they were cooperative, constructive and helpful, or not) should trump all other considerations.[50] SL's insistence that he was *right* and that others should *apologise or else* constituted an almost inverted image of the ideal disputant.

Two days before, in the course of the 2006 project leader election campaign, the candidates were asked when, in their view, did a developer's actions 'cross the line from annoying to destructive' and motivate his removal from the project? The rationale for exclusion was provided by a quote from an IRC channel: '<sl> terry: i hope we never again meet in public, because i promise I will hit you if i do'.[51] Candidates responded that expulsion was the last resort: if no cooperation was possible, this was a sign the project had failed. On 27 April 2006 it was revealed that SL had been expelled from the Debian-Installer (d-i) team. This became SL's central grievance: his ability to upload files directly to the repository ('commit access') had been removed. When it was suggested that he simply send his patches and that others could integrate them, he replied that he refused to be treated like a 'subhuman' by the team: since he had created some of these files, and been maintaining others, nothing less than full privileges would do. This was also a question of honour; the only alternative was forking off part of the code. Soon after, the matter was formally brought before the Technical Committee. The team leader who had removed SL's access apologised for not informing him of this decision, but not for the decision itself, which was considered to have been justified because of past behaviour.[52] On 16 May, the then project leader articulated a central point: it would be perfectly possible for SL to keep working without commit access. What mattered was that 'the consequences of restoring your commit access would be that [the other developers'] contribution would be discredited in so far as they aren't allowed to determine who they work with'. It is not clear what exactly would be 'discredited': autonomy, honour or a combination of both? In any case, the DPL had clearly sided with the team leader and another developer could optimistically write: 'if anything, we now have an official decision (even though you might not like it), so we can all move along'.[53]

No one was going anywhere. Unlike Alegre, SL did not fork. He had, it appeared, nowhere to go. The debate had bounced from list to list: from debian-release to debian-boot, a detour through debian-cttee, and onto debian-vote. SL stayed, and (presumably) stewed, and eventually hit on a strategy: until he got his privileges back, and until those who had wronged him apologised, he would flood the synapses of project, the lists. At least, those lists from which he had not been excluded, chiefly the non-technical ones. The controversy was reignited during the 2007 DPL election. First, SL announced his candidacy, then withdrew it. During the campaign, a developer asked the candidates what their estimation was of how they had handled or would have handled the conflict, and on 16 March SL himself asked a question to the candidates: were the troubles in Debian due to 'arrogance and pride as well as a failure to communicate'? On or around 28 March the Debian account managers suspended his account for a year. He could not upload code; but no one banned him from the lists. On a thread about the newly elected DPL, SL railed against 'Mafioso-like politics in Debian' (29 April). This elicited the following response: 'SL, fuck off. It's not always about you' (5 May). Still, on he went, criticising the manipulative lies of the previous DPL; arguing that his expulsion was illegitimate, a conspiracy by the DAMs, it was a shame for Debian (5 May). Others beseeched him: 'Please, give us some peace. Give yourself some peace. Go and find a new project where you can make a valued and appreciated contribution' (10 May).[54] SL refused, saddened by the fact that there was very little hope that any of his opponents 'would ever be able to do the honourable thing in these actions' (10 May). Debian had a fundamental inability to handle social conflicts; people wanted to hurt the other side as much as they could; a new committee was needed; his expulsion was like a verbal lynch mob; the leader of the d-i team had abused his technical power because of pride and arrogance (11 May); he was a victim sacrificed, and left bleeding on the roadside, because the other side could not accept anything but full bloody victory (18 May). When someone responded that there was 'no shame, only annoyance', SL responded: 'This behaviour is the idea of considering fellow DDs as machines which can be exploited and thrown away at the minor inconvenience. There is shame, and I question your honour and human decency for not recognising it' (23 May).

What is striking about the case is the intrusion of personal and emotional themes into the sphere of code and power. SL's messages were often very personal, questioning his opponents' honesty or their credentials. But he also constantly explained how the actions of others affected *him*, the hurt and suffering *he* felt. He counterposed honour as pride and arrogance to honourable actions in the non-archaic sense of fair and just. By any reasonable measure he was violating all the standards of Usenet and Debian netiquette (cross-posting to multiple lists, flooding lists, trolling) but *he wanted to code*, he would not leave; in his view, he had been treated badly, and a just and fair resolution demanded his reinstatement.

No one was willing to execute the plaintiff socially, to banish him from the lists; the listmasters were reluctant to exercise authority. An outcome of the saga was renewed opportunity to discuss conflict management on the lists. Similar problems occurred elsewhere: an arbitrary decision was made; people accepted it, or left. The difference was that in corporate contexts, people in authority were not shy about making decisions, made them much earlier and faster, and enforced them in a 'considerably more draconian fashion' than Debian did.[55] IJ, the head of the Technical Committee, proposed that a *social committee* should be established, as 'recent events have shown us again that we need an advisory and disciplinary process short of expulsion'.[56] This would help to resolve non-technical disputes. But should this body be formally established by the constitution, or just be delegated by the DPL; would the latter give too much power to the DPL? The social committee proposal was not successful.

A year later, familiar calls were heard: Debian was not working properly! Former DPL AT invoked SL's conflict with another developer, as it was probably the 'most extreme example of a problem Debian had to resolve'.[57] This conflict had been escalated to the DPL, to the Technical Committee, to the DAMs, to the FTP master and others. AT argued that of all these groups, the TC had been the least effective, having been unable even to admit that it was incapable of addressing the issue in a timely manner.[58] The solution was obvious: it was the *Technical Committee* that needed to be reformed. AT suggested that

> replacing the longest serving member with someone else makes it easy to get new ideas and knowledge into the committee, and avoid having an aristocracy/priesthood/whatever of developers who think they're above the laws everyone else abides by, without having to criticise the existing members.[59]

Members of the Technical Committee, all highly skilled and (in some cases) sitting in place since the beginning of Debian, embodied hacker-charismatic authority. Just like its inspiration, the US Supreme Court, there is no revocation mechanism from the project's instance of last resort. In AT's view, sclerosis was an inevitable consequence of this permanence. Not everyone agreed. The project's secretary (also a member of the TC) argued that limiting the number of roles held by developers would not be useful, as the number of roles did not seem to be a good predictor of performance. Someone else then raised an objection: wasn't the secretary, as an 'interested party', affected by this discussion? The secretary's response was *not* humble, declaring this argument to be the 'kind of censoring bullshit that does annoy me'.[60] When the other person questioned the bluntness of his tone, the secretary responded with a how-to guide on defending one's injured honour: 'While you, sir, may think that way, and modify your behaviour to retain a position with additional duties, because you think that means a position of power, I would find that sort of change in behaviour sycophantic, fawning, unethical, and below me'.[61]

There the discussion ended. The defence of honour foreclosed the examination of what is indeed a core problem: the evaluation and acceptance of infrastructural improvements rests with central people occupying central functions. Since these developers are also those most able and confident to make changes in other aspects, they are overworked and cannot train new entrants or document their work. As RA observes, in a corporate environment they would be compelled to set time aside for mentoring and documenting: Debian cannot hope to emulate a corporation in this regard.[62] This issue has been noted in relation to the Debian System Administrators team (DSA), who do not engage in communication, reporting or documentation of changes.[63]

Of all the projects examined here, Debian is by far the most revealing of what tribal distributed leadership would entail for the management of complex infrastructural systems. The examples of Daily Kos and (in the next chapter) Wikipedia are significant, but the stakes are much higher when participants can cause significant harm to the project.

8

THE GREAT SOCK HUNT: wikipedia.org

> 'God save thee, ancient Mariner!
> From the fiends, that plague thee thus! –
> Why look'st thou so ?' – 'With my cross-bow
> I shot the ALBATROSS.
> ...
> Ah ! well a-day! what evil looks
> Had I from old and young !
> Instead of the cross, the Albatross
> About my neck was hung.'
> Samuel Taylor Coleridge, *The Rime of the Ancient Mariner*

When it comes to Wikipedia, the analytical framework used in these case studies, which neatly separates 'project', 'authority' and 'conflict', reaches its limits as these three notions inexorably converge into a single tumultuous entity. Wikipedia rests on the premise that anyone with Internet access can be an expert, and that the combined efforts of this swarm of amateur authorities will necessarily result in correctness. This epistemic authority of the multitude means that each page on Wikipedia is potentially a battleground on which individuals put forward their versions of the truth. Further, the open nature of the wiki software – anyone can edit a page, and the change will be instantly apparent – means that vandalism and self-promotion are constant challenges. In response, a series of behavioural norms ('wikiquette') and a caste of specialised officers ('sysops' or 'admins') have arisen.

This book is mostly concerned with administrative or executive authority (the legitimate power to control others). A frequent source of confusion, which was pointed out in the Introduction, is that in certain online tribes, such as Free and Open Source Software projects, autonomous learned authority or expertise is all-important, and forms the basis of leadership. For participants in non-hacker Internet projects, claims to learned authority are viewed with suspicion. However since the purpose of an encyclopedia is to collate the best of human knowledge, offline expertise informs many Wikipedia debates.

Project: Expert Texpert Choking Smokers

Wikipedia brings together all the issues raised by the distribution of governance and expertise, starting with this distribution's immense appeal. Affording participants maximum autonomy has generated phenomenal development, and this makes it difficult to get a 'handle' on this organically proliferating rhizome.[1] The production of a free knowledge repository makes contributors feel that they are involved in 'something to benefit mankind'.[2] The result of their combined efforts is a Borgesian 'garden of forking paths': a labyrinth of meaning, underlaid by a subterranean layer, the 'talk' or 'meta' pages on which users discuss content and policy.

In the beginning was Nupedia, launched in March 2000 by Jimmy Wales, an Internet entrepreneur, and Larry Sanger, an academic. Like Debian, the intention was for experts to produce top-quality data, so contributors would need to be credentialed: 'We wish editors to be true experts in their field and (with few exceptions) possess Ph.D.s', declared Sanger. But Nupedia grew slowly, and it was its side-project, using Ward Cunningham's wiki software, which survived and flourished: Wikipedia had 17 articles in January 2001, 150 in February, 572 in March, 835 in April, 1,300 in May, 1,700 in June and 2,400 in July.[3] That same month Larry Sanger speculated that, if Wikipedia continued to grow at the rate of 1,000 articles per month, in seven years it would have *84,000 articles*.[4] Seven years later, the English version comprises more than *2 million* entries. Wikipedia has inspired a legion of specialised Internet encyclopedias such as Uncyclopedia, Chickipedia, Wookieepedia, Dickipedia, Dealipedia, Congresspedia, Bulbapedia and Conservapedia.

As on Daily Kos, credentials were to matter less than the tribe's efforts – though not everyone agreed. Larry Sanger, for one, felt that the voice of experts should carry more than that of others, and was soon embroiled in flame wars with the likes of 'The Cunctator'. Eventually the money for Nupedia ran out and Sanger departed from Wikipedia, feeling that he had become a symbol of authority in an anti-authoritarian group.[5] This left Jimmy Wales, commonly known as 'Jimbo', in charge. Without a doubt, Wales occupies a special place in Wikipedia. Semi-facetiously known as the project's 'God-king' or 'benevolent dictator',[6] he is Wikipedia's chief spokesperson and tireless champion. In June 2008 Wales wrote in a British newspaper that an open encyclopedia requires a ruthless precision in thinking because, in contrast to the 'comfortable writers of a classic top-down encyclopedia', people working in open projects are liable to be 'contacted and challenged' if they have 'made a flawed argument or based [their] conclusions on faulty premises'.[7] Central to Wikipedia is the radical redefinition of expertise, which is no longer embodied in a person but in a *process*: the aggregation of many points of view. This, in essence, is the famous concept of the 'wisdom of the crowd' which applies to knowledge the free-software slogan that 'with enough eyeballs, all bugs are shallow'. As an undergraduate, Jimbo had read Friedrich Hayek's free-market manifesto, *The Use of Knowledge in Society* (1945), which argued that people's knowledge is by definition partial: the truth is only established when individuals pool their wisdom.[8] This is why the inclusion of draft articles (known as 'stubs'), no matter how rough, is encouraged in Wikipedia: they can always be edited and become pearls of wisdom. Wales understood what the project needed to succeed: people. The 'wisdom of the crowd' model would only work if the crowd showed up in the first place, and that depended on Wikipedia being an open and friendly environment in which participants would have fun:

> At some ultimate, fundamental level, this is how Wikipedia will be run ... Newcomers are always to be welcomed. There must be no cabal, there must be no elites, there must be no hierarchy or structure which gets in the

way of this openness to newcomers ... 'You can edit this page right now' is a core guiding check on everything that we do. We must respect this principle as sacred ... Anyone with a complaint should be treated with the utmost respect and dignity. Diplomacy consists of combining honesty and politeness. Be honest with me, but don't be mean to me. Don't misrepresent my views for your own political ends, and I'll treat you the same way.[9]

With input from the community, Sanger had written definitions, such as 'What Wikipedia is not'.[10] These writings formed the so-called five pillars on which the project is built. (1) Wikipedia is an encyclopedia: no original research is allowed. (2) Contributors must adopt a 'neutral point of view' and use verifiable, authoritative sources. (3) The articles are free content that anyone can edit, as the text is available under the GNU Free Documentation License. (4) Contributors should follow the Wikipedia code of conduct, being civil, finding consensus, and avoiding 'edit wars' (editorial-content disputes). (5) Finally, there are no other firm rules aside from these five general principles.[11] In fact, taking part in Wikipedia in a sustained manner requires a commitment to 'a style of writing and describing concepts that are far from intuitive or natural'.[12] Adhering to editorial and behavioural policies and guidelines is essential on Wikipedia.

The 'wisdom of the crowd' development model holds that adding contributors makes for better content, and this has been empirically borne out: the rigour and diversity of a Wikipedia article improve following a reference to this article in the mass media, which brings in new contributors.[13] Editors can also arrange for articles to appear on the front page in the 'Did You Know' section (WP:DYK) which lists new additions to the encyclopedia. This can attract between ten and twenty editors, who not only work on the article but also put it in their watchlist.[14] In the early days articles were copied from public-domain sources such as the 1909 edition of *Encyclopaedia Britannica*.[15] But for the most part, online encyclopedists find their sources of information on the Internet. Wikipedians are concerned with *verifiability* rather than truth, and the Internet is a handy way of cross-checking information and sources. As a result, a pattern of mutual justification between Wikipedia and the Internet's index authority-attributing mechanism, otherwise known as Google, has come into being. Wales once declared: 'If it isn't on Google, it doesn't exist.'[16] One problem with this approach is that things which have *not* been digitised, such as underground or marginal culture, are then likely to be deleted because they are non-verifiable online. Another problem is that Wikipedia is itself returned in the top five searches of Google searches, because Wikipedia pages contain many links to other pages on the site, and are frequently updated. Critics suggest this results in a confusion between *authority* and *popularity* so that quality, reputation and expertise become conflated.[17] In other words, there simply is no guarantee that the many eyes of the crowd will fix all the bugs. Jaron Lanier, in an article entitled 'Digital Maoism', derides the 'fallacy of the infallible collective' manifest in a wiki or 'other Meta aggregation rituals', which results in the loss of discernment, taste and nuance.[18] It is certainly true that user-moderated news aggregator sites such as Digg create averages which offer no guarantee of correctness, only of popularity amongst people who use the software.

The Wikipedia model came under serious attack for the first time after the 'Seigenthaler controversy'. On 26 May 2005, while at work at Rush Delivery in Nashville, Brian Chase created a biographical entry on John Seigenthaler, a journalist, writer and former Robert Kennedy aide. In order to amuse a colleague, Chase wrote that Seigenthaler was 'directly involved' in the assassinations of John and Robert Kennedy, had moved to the Soviet Union in 1971, returned to the United States in 1984, and 'started one of the country's largest public relations firms shortly thereafter'. The new pages patrol, in charge of monitoring additions to the encyclopedia, was unable to keep up with the flow of new articles and the hoax was not discovered until September. In October, Seigenthaler contacted Wales, who removed the hoaxed versions of the article's history from the Wikipedia version logs. New page creation by anonymous editors was disabled in November 2005. On 29 November 2005, Seigenthaler wrote about the incident in *USA Today* and called Wikipedia a 'flawed and irresponsible research tool'.[19] After the case was publicised many commentators on the article's talk section were unsympathetic towards the victim, repeatedly asking why such a free-speech advocate was threatening Wikipedia, and why Seigenthaler had not simply fixed the misinformation himself.

Passing off the responsibility onto the user for dealing with the inadequacies of software or information may be an adequate response when people *choose* to participate in a project, as is the case with free-software projects, for example. It is manifestly inappropriate in the context of an encyclopedia, where people have no say as to whether they are being written about or not. These contradictions have caused Wikipedia to be violently criticised. An anonymous commenter to an online magazine article likened Wikipedians to critics of the theory of evolution, and lambasted the 'popular movement to sneer at "experts" and "academics" as if they've been oppressing us all through history with their malicious insistence on evidence'.[20]

Mendacious modifications can also be motivated by self-interest. In 2007 Virgil Griffith created Wiki Scanner, a program that identified which organisations had been editing Wikipedia articles. Transparent cases of self-promotion were revealed: someone accessing Wikipedia from the IP address of the voting-machine company Diebold had apparently deleted long paragraphs examining 'the security industry's concerns over the integrity of their voting machines' as well as information about the company's CEO raising money for President Bush.[21] The strongest opposition to Wikipedia has come from the ranks of *Encyclopaedia Britannica*, whose corporate brand, established over time, was naturally threatened by the free wiki. A comparison in the scientific journal *Nature* of 42 hard science articles in Wikipedia and *Encyclopaedia Britannica* found that their quality was roughly similar.[22] *Britannica* subsequently disputed the study's methods and findings as biased and unscientific.[23] The *Britannica* perspective is that in all cases of user-generated content – whether an Amazon review or a Wikipedia article – the *quality* of the eyes examining a project trumps their *quantity*.

Other encyclopedic user-generated projects have emerged with the stated aim of improving the reliability of articles. Google's knol project purports to reinforce authority by using recognised experts, but plans to introduce sponsored links next to encyclopedic articles. Money would be paid back to authors of articles, but it is unclear what credit should be given to such sponsored knowledge. Citizendium is Larry Sanger's attempt to marry collaborative editing with expert contributors and a credible review process. Finally, some Wikipedians have created Veropedia, which 'freezes' quality Wikipedia articles (they cannot be edited), resulting in 'a stable version that can be trusted by students, teachers, and anyone else who is looking for top-notch, reliable information'.[24]

Authority: The Cabal

Authority in Wikipedia is applied to articles and people. Editorial authority is in principle open to everyone; further, articles are unsigned and produced by countless authors. Nonetheless, contributors refer to authorship information when monitoring edits to Wikipedia: they are more suspicious of edits made by anonymous or new users than of those made by editors with already established records of valuable contributions, for example.[25] Anonymous users are identified by their IP addresses.[26]

Contributions to the project are statistically measurable by software tools: in the first instance, reputation is a function of the number of edits.[27] In order to drive up their edits counts, editors may be tempted to make many small edits rather than 'big picture' or 'high content' edits. Registering as an editor automatically generates a 'user page' where quantitative metrics such as number of edits are displayed alongside more subjective ones. Editors list articles they have initiated or significantly contributed to, often detailing whether these articles were distinguished in some manner ('featured articles', for example, appear on the encyclopedia's front page for 24 hours). Editors can also list their membership of various 'wikiprojects' (groups of articles classified by genre, and other collective projects); their administrative responsibilities in the project; the accolades they have received from other Wikipedians in the form of five-pointed 'barnstars' – e.g. user X awards user Z a 'random acts of kindness barnstar' or a 'tireless contributor barnstar'.

What distinguishes Wikipedians from outsiders is their familiarity with project language and rules. Wiki-vocabulary includes 'forum shopping' (canvassing for support), 'fancruft' (unencyclopedic content), 'smerge' (small merge), 'hatnote' ('short notes placed at the top of an article before the primary topic') and the like. This specialised language does not appear in 'article space' but in talk pages. Talk pages are editable discussion spaces where, in contrast with articles, messages are signed. Every user page, article page and policy page has a talk page; some talk pages have more than 50 pages of archives. Participants use talk pages to resolve disputes but also as a site for collective planning, with frequent requests for coordination and information. They also serve to socialise newcomers via references to policy guidelines. A crucial sign of competence in Wikipedia is the capacity to authoritatively assert that edits are in accordance with project policy and guidelines. As of September 2007, there were 44 pages in the 'Wikipedia official policy' category; 248 pages were categorised as 'Wikipedia guidelines', organised into eight subcategories; and, if that were not enough, 45 proposals for guidelines and policies were pending.[28] Viegas et al. suggest that it is because Wikipedia's records are persistent, public and easily available online that norms and guidelines keep being created.[29] As we shall see in the next section, reference to policy is the major rhetorical weapon employed during edit wars. Some policies are meant to guide the quality of articles, which should have a neutral point of view (WP: NPOV), be verifiable (WP:V) and not constitute original research (WP:OR). Some regulate relations with other contributors: editors should assume good faith (WP:AGF), not revert someone else's edit more than three times in 24 hours (WP:3RR) and avoid personal attacks (WP:NPA).

Can people pull rank in a rankless universe? The invocation of outside competences or credentials contradicts the 'wisdom of the crowd', but it happens. After the Seigenthaler incident, the episode which did the most damage to the project's credibility was the case of Essjay, a Wikipedia editor who was so well regarded that he was put forward to be interviewed for an in-depth profile of Wikipedia by the *New Yorker*. He was also offered a job as 'community manager' with Wikia, the for-profit organisation launched by Wales in 2004. This was to prove his undoing, for the biographical information he included on his Wikia pages did not match the profile Wikipedians or *New Yorker* readers had been given: it turned out that Essjay was not in fact a professor of divinity with a BA and an MA in religious studies and doctorates in law and philosophy, but a 24-year-old with no academic credentials whatsoever. Essjay had repeatedly used these bogus credentials to bolster his views during content conflicts with others. In a discussion over the term 'imprimatur' as used in Catholicism, Essjay defended his use of the book *Catholicism for Dummies* by declaring: 'This is a text I often require for my students, and I would hang my own Ph.D. on its credibility'.[30] After asserting that he had used the deception to protect himself from the unwanted attention of 'trolls, stalkers and psychopaths' Essjay was eventually asked to leave Wikipedia and Wikia by Wales. What made the case so jarring was that Essjay's otherwise genuinely helpful dispute-resolution and counselling work had led him to the heights of Wikipedia's governance structure: he was a *mediator*, a *sysop* (or *admin*) with *oversight* and *CheckUser* privileges, a *bureaucrat*, and a member of the *ArbCom*. Once his identity had been revealed he joked on his user page on 6 February 2007: 'One nice thing about being "out" is that now I get to hang out with the rest of the cabal in real life; I had dinner tonight with Jimbo, Angela, Datrio and Michael Davis.'[31] The existence of a cabal – a shadowy inner circle secretly controlling everything – has been an Internet joke since the days of Usenet. Thus we come to the question of administrative power. On Wikipedia, ordinary editors wield power over *article components*, but administrators wield power over *editors* and *articles*. The clearest manifestation of administrative power on a digital network is the capacity to exclude participants, or to strip them of some of their privileges. Originally Wales dealt with every instance of vandalism, but a raft of roles and procedures was progressively created. In October 2001, Wales appointed a small group of system administrators.[32] The rising volume of contributions eventually compelled Wales to formally announce in 2003:

> I just wanted to say that becoming a sysop is *not a big deal*. I think perhaps I'll go through semi-willy-nilly and make a bunch of people who have been around for awhile sysops. I want to dispel the aura of 'authority' around the position. It's merely a technical matter that the powers given to sysops are not given out to everyone. I don't like that there's the apparent feeling here that being granted sysop status is a really special thing.[33]

Content creators are usually 'pre-admins': they are primarily occasional editors, specialists. A study analysing the work of a sample of Wikipedia editors showed that new users created three-quarters of the high-quality content, especially during their first three or four months on-wiki. Initially admins produce more content at a less rapid pace, but as they become more involved in meta-matters their contributions become both more frequent and less content-oriented.[34] Their primary concern is now for the health of the project itself; they have become custodians. This division between content-oriented and process-oriented users can cause tensions. In 2007, a proposal ('prise de décision' or PdD) defining the use of scientific terminology or vernacular language for the classification of zoological species on the French Wikipedia generated a rancorous debate. The objections to the proposal were that it was not procedurally sound, and it was ultimately defeated. One of the proposal's authors took a 'wikibreak' to calm down. Returning to the project two weeks later, she wrote on the administrator's noticeboard about her feeling of unease when she realised that most opponents of the decision had less than 40 per cent participation in the encyclopedic part of the project (one having less than 10 per cent), whereas most of those who had initiated and supported the proposal had participation rates in the encyclopedia which were higher than 80 per cent. There were people, she realised, who *specialised* in pages where votes were held.[35]

Editors nominated for a request for adminship (WP:RFA) must field questions from the community for seven days in order to assess their experience and trustworthiness. Close attention is paid to a candidate's record on handling contentious issues, such as content disputes with other editors. Any registered user can ask questions or vote. As with most decisions on Wikipedia, the decision is not based on strict numerical data but on 'rough consensus' (as determined by a bureaucrat), which means receiving around 75 per cent of support.[36] It is proving increasingly hard to become a Wikipedia administrator: 2,700 candidates were nominated between 2001 and 2008, with a success rate of 53 per cent. The rate has dropped from 75.5 per cent until 2005 to 42 per cent in 2006 and 2007. Article contribution was not a strong predictor of success. The most successful candidates are those who have edited the Wikipedia policy or project space; such an edit is worth ten article edits. Conversely, edits to Arbitration or Mediation Committee pages, or to a wikiquette noticeboard, decrease the likelihood of being selected.[37]

What powers do admins have? They can grant *rollback* (accelerated reversion) privileges to ordinary editors. They can also delete and undelete articles, protect and unprotect them from further changes, and block and unblock users from editing. They can also take on additional responsibilities. *Bureaucrats* can grant administrator status, rename users, and grant bureaucrat status. *Stewards* can change all user access levels on all Wikimedia project. *Developers* can access MediaWiki software and Foundation servers. In addition, two special software procedures are *oversight*, which enables the hiding of page revisions from all other user types, and *CheckUser*, which uncovers the (normally hidden) IP addresses of registered users. As the volume of work and of disputes grew, a Mediation Committee aiming to find common ground between edit warriors was established; it had no coercive power. Eventually Wales decided to establish an Arbitration Committee (ArbCom) which could impose solutions that he would consider binding, though he reserved the right of executive clemency and indeed even to dissolve the whole thing if it turns out to be a disaster. But I regard that as unlikely, and I plan to do it about as often as the Queen of England dissolves Parliament against their wishes, *i.e.*, basically never, but it is one last safety valve for our values.[38]

The Arbitration Committee only comprises a dozen individuals; in effect, it constitutes Wikipedia's Supreme Court, being the last step in the dispute resolution process.[39] Despite these trappings of democratic rule, the fact remains that, following advisory community votes, successful ArbCom candidates are *appointed by Wales* for three-year terms. This is just one example of the manner in which Wikipedia exhibits the opposite pull of the two principle forms of legitimate power in online tribes, charismatic and sovereign authority.

Leadership on Wikipedia is intimately associated with the person and beliefs of Jimmy Wales. His Wikipedia entry tells us that he was a follower of Ayn Rand, running the mailing list 'Moderated Discussion of Objectivist Philosophy' between 1992 and 1996. A libertarian, then, whose guiding principles are freedom and liberty in the sense of 'not initiating force against other people'. In 2006 Marshall Poe approvingly described Wales's 'benign rule', asserting that Wales had repeatedly demonstrated an 'astounding reluctance to use his power, even when the community begged him to', refusing to exile disruptive users or erase offensive material.[40] However Wales, like the Wizard in the MUD, still wields extraordinary powers. User:ZScout370 contradicted Wales by unblocking a block of a problematic user made by Wales: Jimbo slapped a week-long ban on him.[41] In July 2008 Wales intervened in a discussion about whether an admin had acted appropriately when accused of misogyny by stepping in and cursorily 'desysopping' the admin.[42] Since they were performed by the project's charismatic founder, these actions were not necessarily perceived as illegitimate. But they contradicted the procedural basis of a sovereign authority regime, and generated controversy.

The crucial fact about Wikipedia's distributed authority mechanisms is: *there are more and more of them.* A study by Kittur et al. found that non-encyclopedic work, such as 'discussion, procedure, user coordination, and maintenance activity (such as reverts and vandalism)' is on the rise.[43] Conversely, the amount of direct work going into edits is decreasing: the percentage of edits made to article pages has decreased from over 90 per cent in 2001 to roughly 70 per cent in July 2006, whilst over the same period the proportion of edits going towards policy and procedure has gone from 2 to 10 per cent.[44]

What does Wikipedia's phenomenal growth and the inflation of procedure-oriented pages mean for social relations within the project? Benkler and Nissenbaum argue that Wikipedia constitutes a remarkable example of self-generated policing. They extol the project's use of open discourse and consensus as well as its reliance on 'social norms and user-run mediation and arbitration rather than mechanical control of behaviour'.[45] This ideal scenario can be contrasted with Bryant et al.'s observation that Wikipedia software is designed to *encourage the surveillance of others' contributions*, through watchlists for example.[46] The proper formatting of the untamed energy of the crowd is indeed the central dynamic of Wikipedia. The overwhelming majority of new policies and rules apply to editors, who need to be controlled, not to admins.[47] The exponential growth in the number of participants has resulted in admins taking on roles that are more social than technical. A series of interviews with editors at varying levels of authority found that almost all the interviewees believed that 'the role of administrator carries with it more social authority than it ever has in the past'.[48] Whereas admins previously relied on community consensus or on the ArbCom to decide article protections and deletions and user exclusions, they are in effect becoming interpreters of policy – judge, jury and executioner. This in turn has resulted in two major consequences: as we have seen, the process of becoming an admin has become increasingly more arduous. Another consequence is increased debate about the role of admins, and in particular about potential abuses of power. This is the focus of the next section.

Conflict: The Durova Dust-up

Contributors to Wikipedia often exhibit a genuine spirit of cooperation and consensus, suggesting compromises and thanking one another for help in resolving problems. These achievements notwithstanding, and despite exhortations to assume good faith, stay cool, embrace 'wikilove', observe wikiquette, and forgive and forget, they also frequently fight like cats and dogs. This is to some degree unavoidable, given that the project rests on foundational principles such as 'notability', which are open to subjective interpretation. Where editors position themselves in relation to the central editorial decision of whether an article should be deleted has been formalised as a philosophical distinction between *inclusionists* who do not wish to exclude information and *deletionists* who advocate following strict guidelines on what is encylopedic.

Since an influx of knowledgeable newbies is crucial for generating new content, a special recommendation made to editors is that they should refrain from 'biting' (being mean to) newcomers. A precise set of guidelines explains how to deal with people who may not be attuned to the finer nuances of Wikipedia formatting and etiquette.[49] In general terms, edit wars erupt over controversial topics. Aside from the perennials – Northern Ireland and Palestine – any area of human activity about which people have strongly contrasting opinions, typically but not exclusively involving international, interfaith or inter-ethnic conflict, are potentially fertile ground for 'wikidrama' and edit warring. Is the Korean martial art tae kwon do primarily a reworking of Japanese karate or does it incorporate home-grown elements? Is Turkey part of Europe, the Middle East or Asia? Should John Howard's page mention that some people want him tried before the ICC for his role in the invasion of Iraq? Disagreeing is part and parcel of the experience, and the constant back and forth should ideally reach the point of consensus, at which deliberation delivers its rational result. All too often, however, it is the most persistent debaters who succeed in wearing down their opponents' resistance.[50]

The easiest way to defeat an opponent is to assert that their views are not authoritatively backed up by a proper source, that they are violating the sacrosanct WP:NPOV (neutral point of view) or WP: RS (reliable sources) rules. By extension, all references to editorial, stylistic and behavioural policies and guidelines serve as weapons in battle. It sometimes seems as if every single action having to do with the project has been distilled into a convenient WP: SLOGAN, ready to be whipped out at the slightest provocation. Furthermore, when debating or appealing to others, editors are expected to provide links to evidence, or DIFFS. DIFFS are pages showing the difference between two versions of a page, which are automatically generated and archived each time an edit is made to a page. Fighting also occurs between admins. A 'wheel war' is a struggle between two or more administrators in which they undo each other's administrative actions – unblocking and reblocking a user, undeleting and redeleting, or unprotecting and reprotecting a page.[51] What are users to do if they feel that someone has breached a policy, not participated in discussion, or unprotected a page which has consistently been targeted by a vandal? There exists a veritable phalanx of pages and noticeboards designed to request advice and assistance, report abuse and discuss problems.[52]

Since anyone can contribute, the temptation to cause mischief is seldom resisted. There are many shades of vandalism, including 'misinformation, mass deletions, partial deletions, offensive statements, spam (including links), nonsense and other'.[53] Widespread vandalism has resulted in the emergence of a new breed of sysop, whose main claim to adminship is their work as 'vandal bashers', using reverting software such as Huggle. Defacements occurring in 'articlespace' are easily detectable and reversible, especially when they are crude or juvenile. More insidious vandals attempt to abuse the policing system. The deliberate misuse of administrative processes is a favourite troll game. Examples include the continual nomination for deletion of articles that are obviously encyclopedic, the nomination of stubs (draft articles) as featured-article candidates, the baseless listing of users at WP:Requests for comment (a dispute resolution mechanism), the nomination of users who obviously do not fulfil the minimum requirements at WP:Requests for adminship, the 'correction' of points that are already in conformance with the manual of style, and giving repeated vandalism warnings to innocent users.[54]

Aside from purely malicious vandalism, several kinds of 'illegal' participation occur on Wikipedia. A currently undetectable form of 'wikicrime' uses a 'tag team' approach. Participants manipulate the system by forming cliques to get an opponent reprimanded or sanctioned. Tag team-mates coordinate their actions by email or IRC to take turns in taunting or opposing their opponents, thereby goading them into breaking an important rule, such as the 3-Revert rule. The most notorious crime on Wikipedia however is the use of 'sockpuppets' (known in the French version as *faux-nez* or 'fake noses'). These enemies within the project arise when people create alternative accounts, in addition to their existing Wikipedia identities, in order to take part in debates and votes. 'Socking' perverts the search for consensus; sometimes 'socks' are created so that users can conduct arguments with themselves. Some editors have been found to have created hundreds of fake personas. Convicted sock-puppeteers are banned on sight, but how can one tell if a sock is at work? Certain signs are telling: socks exhibit a strong interest in the same articles as their other personae; they often employ similar stylistic devices; and they make similar claims or requests as their puppet master. When editors suspect that a user is exhibiting 'sockish' behaviour, or that a 'sock farm' has been detected, they can call on a special weapon to be used. This is the CheckUser software, accessible to a restricted number of admins. CheckUser identifies what IP address registered Wikipedians are accessing the site from. If it is found that distinct accounts involved in disputes are issuing from the same terminal, Wikipedia's authorities can ban entire areas or ISPs. Though technology-savvy users can always use proxys or anonymising software such as Tor, CheckUser is seen as a valid means to identifying vandals, and those admins who are entrusted with it are held in high regard.

Disagreements on Wikipedia have a habit of finding their way onto other websites and sometimes into the offline world, leading to charges of harassment and stalking. 'Watchdog websites', such as Wikipedia Review, Wikipedia Watch, Wikitruth and Encyclopaedia Dramatica, have sprung up, ostensibly to document the misdeeds of the admin class. These sites have variously disclosed personal details (including photographs) of Wikipedia editors, alleged that some prominent editors are plants of secret service agencies, denounced various forms of bias, and published deleted articles. The latest events on Wikipedia are seized upon, analysed and, more often than not, mocked and held up as evidence that, yes, the cabal has struck again. Accordingly, these sites have been labelled 'attack sites' by Wikipedians.[55]

Beyond these sometimes extreme cases, the fact is that Wikipedia's editorial process, understood as the herding or disciplining of autonomous content providers, cannot but generate bad blood, in the shape of participants who feel they have been ill treated, or even humiliated, by editors and admins. Unfairness or injustice can be hard to evaluate, as both sides in disputes invariably feel that they are in the right, so a structural example will best illustrate the issue: creators of articles set its tone. Because of a 'first-mover advantage', the initial text of an article tends to survive longer and suffer less modification than later contributions to the same article.[56] It is to be expected that article creators who maintain an interest in the article would put it on their watchlist and, despite the project's injunctions, would experience feelings – if not of ownership – at least of heightened sensitivity and possibly unhappiness if someone attempts to 'improve' their baby. The problem is compounded when editors have access to administrative tools.

Critics contend that certain admins engage in 'drive-by reversing' so as to drive up their edit count. An anonymous contributor to a Slashdot discussion about Wikipedia wrote that a series of rapid-fire reversions of article summaries appeared suspect because the reversions had been 'made so rapidly, at a rate of several a minute [that] it is unlikely that the admin was really reading the articles in question'.[57] Critics also allege that Wikipedia is controlled by 'cliques' who manipulate the system for their own biased purposes. Anyone who dares to disagree, charge these critics, is accused of 'wikilawyering', of violating consensus, and is labelled a troll. An ex-editor asserted that expressing his point of view in a message to the Wikipedia English-language mailing list was answered not with an examination of his case, but with 'platitudes about rules and regulations the newcomer did not follow'. Questioning the sagacity of an admin, continued the ex-editor, generated the response: '"You don't get anywhere by attacking an admin" – not even if they were wrong.' According to this ex-editor, Wikipedia adminship has a 'dirty secret': it is a 'cult, a good old boys network, a Masonic society of sorts'.[58]

The feeling of injustice, the notion that the liberatory promise of the project has failed, is at the centre of the conflicts examined in the case studies presented in this book. Until now, the issue has been examined mostly from the perspective of the victims: those who were troll-rated or excluded, and whose feelings of injury and injustice led them to embark on a restorative campaign. In the case of Wikipedia, I now consider the *enforcement* side of the equation. User:Durova was a well-respected admin, the recipient of many a barnstar, and a self-described 'sleuth' who specialised in sock-hunting. By early November 2007 her wiki-career was going exceedingly well, as she was being interviewed on-wiki to assess her suitability to join the ArbCom, Wikipedia's supreme court. She declared on that occasion that she would like to spend time in 'mainspace', but

> it turns out I'm one of a limited number of people who has talent and experience at sleuthing and I'm not easily baited. When I compare how much good it does the project to write one featured article against how much harm it prevents to foil a persistent banned sockpuppeteer, I usually opt for the latter.[59]

On 18 November 2007, Durova blocked User:!!, with an expiry time of 'indefinite'. She advised the admin noticeboard of the fact and provided her rationale: 'Abusing sock puppet account'. Other well-respected admins did not understand the block: there was nothing wrong with User:!!'s edits. Admittedly, !!'s knowledge of project rules and procedures was improbably extensive for a new user. But he or she had made a positive contribution, with more than 100 contributions to WP:DYK. Facing repeated queries, Durova stuck to her story: she could not explain why she believed User:!! was a sock, because opening the door a little bit would then become a wedge for people to ask for more information, and this would compromise some 'rather deep research techniques'.[60] Any questions should be taken to the ArbCom, she added. Eventually someone privately apprised Durova of the facts regarding User:!!, and the truth was soon made public: User:!! was no vandal, but a returning user who had simply wished to change his or her name because of past privacy concerns. So, after 75 minutes, Durova reversed her decision:

=== Unblock with apologies ===
When I make a mistake I like to be the first to step forward to correct myself. It's very surprising that a few facts didn't come to light sooner, given the amount of time my report circulated and the people who had access to it.[61]

Report? What report? wondered Wikipedia. And also: who were the people of importance who had seen the report? And finally: how had Durova communicated with them? It eventually came to light that Durova had, on 3 November, posted a 'lesson in spotting returning editors' to a hitherto secret 'cyberstalking' email list. The email in question was posted by a Wikipedian to his talk page and 'instinctively' oversighted (deleted and scrubbed from the records) by ArbCom member User:Blnguye, an error for which he later apologised.[62] Naturally, copies of the email were posted on 'watchdog' sites: User:!!, had written Durova, was a 'ripened sock with a padded history of redirects, minor edits and some DYK work'; he or she had also engaged in 'obscene trolling in German, and a free range of sarcasm and troublemaking'. In short he or she was clearly not a first account, was too helpful to fellow editors, and worked too hard on mundane tasks.[63] Durova's 'lesson' elicited 'little attention and no response' on the secret list.[64] In reality, the 'obscene trolling' was a reference to a well-worn German phrase and the 'troublemaking' was a veiled suggestion that User:!! might have been a returning user. Durova at any rate had felt that she was justified in proceeding; it remains unclear whether she received encouragement from others. When these facts were revealed, Wikipedia's 'talkspace' erupted. The controversy instantly teleported to a host of new pages: apart from the original admin noticeboard, a new administrator's discussion page was created, as well as a 'request for comment' page where User:!! and Durova were asked to explain their actions. Heated discussions occurred on Durova and User:!!'s talk pages, and on the wiki-en mailing list; and the ArbCom thread in which questions to Durova about her candidacy had been slowly winding down was suddenly kicked back to life. A post to the wiki-en email list crystallised the feelings of many Wikipedians: 'For all intents and purposes, you have exercised your right to vanish', wrote User:Risker. But what if this person then decided to return under a new name, allowing them to continue writing fine contributions? This is why User:!! had created a new account, argued User: Risker: 'You behave as a highly motivated, positive, and valuable contributor to the project. There is absolutely no logical reason for anyone to have concerns about your participation.'[65]

Eventually the whole matter was formally brought before the ArbCom, at which point Durova gave a full justification for her actions. She declared that she had made a good faith error and reversed her decision with apologies after 75 minutes. In her defence, she asserted that a group of around 20 people, most of whom were justly banned from Wikipedia, had banded together to disrupt the project. These people, who shared tactics and operated together, employed a range of stratagems:

> Sometimes they use throwaway sockpuppets like the ones in those diffs I supplied. They also build up some long term socks with padded edit histories that attempt to mimic the behavior of productive Wikipedians. More on that in a moment. Sometimes they persuade legitimate Wikipedians to compromise their integrity ... I assembled a seven point checklist for distinguishing their behaviour from the behaviour of regular Wikipedians. Naturally there would be positive contributions on this type of account – and a lot of them: that's how those people inoculate their socks against blocking and banning.[66]

Like a character in a John le Carré Cold War thriller, Durova was convinced that a deep-penetration double agent or 'mole' was boring deep into Wikipedia. In wiki-speak, a sleeper sock was being built up for adminship. Once this basic premise had been established, all that was good about a contributor's work simply reinforced the notion that they were bad – because it was all 'cover'. On 26 November, Durova resigned her sysop privileges. On 1 December, the ArbCom admonished her to exercise more care in blocking users.[67] As for User:!!, he or she stopped editing.[68] Jimbo Wales had offered comforting words: a mistake had been made, then fixed, so everyone *relax* already.[69] But the damage had been done. User:!!'s ban was having a chilling effect: if an exemplary editor got banned, how would others who were lesser-known, or had no friends, fare against a pre-emptive purge?

Once again, since much communication occurred via private email or IRC it is impossible to reconstruct with any degree of certainty what unfolded. What can be gleaned from the publicly archived talk pages and email discussions is that the incident resonated deeply with many editors, because it commingled *authority* and *secrecy*: tools were being abused to exclude users and possibly to suppress evidence; decisions were being made on secret lists where the accused could not defend themselves; some sysops had been operating as rogue supercops, contradicting what was supposed to distinguish Wikipedia from Britannica: its openness and transparency. Truth-seeking was being used not to build an encyclopedia but to root out evil. Yet the Durova dust-up brought to light an equally powerful, and opposite, feeling: that some admins had been the victims of harassment and stalking because of their work for the project; that these experiences were frightening and painful; and that most of the victims were female.

A sour postscript to the affair was provided when Larry Sanger, Wikipedia's former chief editor, posted a message to the wiki-en list: 'I saw this unfortunate article' (describing the Durova incident), wrote Sanger, who explained that he had felt inspired to reach out to the Wikipedia community in order to 'invite those of you who are seriously disaffected to give the Citizendium model another look'.[70] Sanger outlined the multiple ways in which such an incident could never take place in Citizendium, which had a clear governance model. Citizendium was to be 'an open, representative republic' with an editorial council (a 'legislative'), a constabulary (a 'police force'), an executive committee (an 'executive') and it would soon feature an independent judiciary.[71] Sanger concluded with an apparent dig at Wales: he (Sanger) did not intend to stay at the helm, as it was obvious to him that 'the leader of an allegedly democratic project should actually "step aside" when he's handed over the reins of power'.[72]

How should problematic powerful people be dealt with in autonomous projects? The issue of the impeachment of authority figures has been raised on multiple occasions on Wikipedia. In 2008 User:Amerique put forward an admin recall proposal which was extensively discussed and tweaked on his talk page.[73] The proposal attracted the attention of Jimbo Wales, who commented that any such processes were matters of deep concern, because 'people in positions of trust (the ArbCom for example) [should be] significantly independent of day-to-day wiki politics'. Since Wales was unaware of any cases in which a recall process had been needed, he viewed it as a form of 'process-creep'. If there really were such an example, then the project should simply 'look harder at what went wrong'.[74] This approach to governance – keep it fluid, keep it personal, seek consensus – eschews rigid 'institutional' solution. Its long-term viability is open to question. Since Wikipedia operates following the constant reform and refinement of social norms within the tribe, the issue of changing policy with an ever-increasing number of participants becomes more complex.[75] Forte and Bruckman argue that Wikipedia's lack of a central policy-making system means that 'site-wide policy-making has slowed and mechanisms that support the creation and improvement of guidelines have become increasingly decentralised'.[76] Wikipedia's lack of a constitution, or of clearly defined voting procedures that would enable this constitution to be updated, means there is a danger of the project fragmenting into a multitude of smaller wikiprojects – local jurisdictions over which a limited number of participants will have a say, and who may start writing rules that conflict with others. All these points raise the question of the organisational format of collaborative online projects. This is the focus of my final chapter.

9

ONLINE TRIBAL BUREAUCRACY

The spot where we intend to fight must not be made known; for then the enemy will have to prepare against a possible attack at several different points; and his forces being thus distributed in many directions, the numbers we shall have to face at any given point will be proportionately few.

Sun Tzu

This book started with a simple premise: that it is not enough to say that the Internet, because it is a 'horizontal' or 'many-to-many' system, abolishes authority. New forms of domination have emerged on the Web. Leaders in online projects must *justify* actions such as deciding which patch, comment or modification will be integrated into the code, thread or article. Weber's authority ideal types helped to define the legitimate forms of power at work in the anti-authoritarian online environment: an individual axis of technology-mediated charismatic brilliance or network location, and a collective axis of democratic sovereignty. As shown in Chapter 4, online projects may be positioned on different parts of an online authority quadrant according to their dominant form of authority. One issue unaddressed so far is that of the movement between quadrants. The sector in which radical actors such as primitivists reject the very possibility of legitimate power is a priori not concerned by this question. The major question then is the evolution of authority from online-charismatic to online-sovereign regimes. Nicolas Auray has argued that every time an online project becomes more popular, it generates fewer forks and motivates an inner cabal, which leads to more formal organisation.[1] In the case of Debian it was the perceived authoritarianism of the second project leader which motivated participants to depose him, in effect, and draft a constitution.[2] Debian represents a successful example of a transition to a sovereign form of authority, whose constitutional basis is not questioned. In the other cases studied, the transition was either interrupted, or did not occur. Wikipedia exists in an intermediary state between charismatic and sovereign regimes. As long as its founder maintains extraordinary powers, and formal recall and constitutional procedures do not exist, the project cannot complete its transition to a democratic polity. In the case of Daily Kos, evolution is impossible: weblogs are characterised by an innate imbalance in capacities between bloggers running their own fiefdoms and those who they suffer to comment on their sites.

Until now online authority has been framed in terms of its relation to autonomy and as a resource that individuals draw on in order to legitimately administer and police projects. But what is the institutional context of these actions? Remixing Weber, I contend that legitimate power in cooperative online projects is the foundation of a new organisational arrangement, *online tribal bureaucracy*. After specifying how online tribal bureaucracies differ from other organisational types, this chapter focuses on the role of conflict within this organisational form and finally examines their wider political and economic implications.

Organisation Without Domination

So far, terms such as *tribe, community*, or *project* have been used to characterise the online groups examined in this book. Now that we have a clearer understanding of how individuals justify their use of power in online contexts, it is time to analyse the organisation this online authority legitimises. The governance structures of anti-authoritarian groups have been neglected by researchers, compared to more established or conventional organisations. However, earlier conceptualisations do exist. In 1960, Harrison described decentralised church groups as *volunteer associations*. He argued that such anti-authoritarian structures are always unstable because of their contrary requirements:

they require a bureaucracy to realise the mission of their organisations [but] their ideology is rooted in democratic traditions which are inimical to the tendency of technical bureaucracy to depersonalise the individual and to segregate roles on a functional and hierarchical basis.[3]

Following the creation of free medical clinics, legal collectives, food cooperatives, free schools, and alternative newspapers in the 1970s, Rothschild-Whitt defined *collectivist organisations* as alternative institutions which 'self-consciously reject the norms of rational-bureaucracy'.[4] Aside from their value-rational orientation to social action (based on belief in the justness of a cause), collectivist organisations are groups in which authority resides not in the individual, by virtue of incumbency in office or expertise, but 'in the collectivity as a whole'; decisions become authoritative to the extent that all members have the right to full and equal participation. There are no established rules of order, no formal motions and amendments, no votes, but instead a 'consensus process in which all members participate in the *collective* formulation of problems and *negotiation of decisions*'.[5] If this sounds eerily similar to online projects, it's because it is. The Internet Engineering Task Force always took pains to portray itself as anti-bureaucratic, as a collection of 'happenings' with no board of directors or formal membership, and open registration and attendance at meetings: 'The closest thing there is to being an IETF member is being on the IETF or working group mailing lists.'[6] The actual publication of the third request for comment (RFC) performatively established the IETF's collaborative process: there would be a working group of implementers actually discussing and trying things out; ideas were to be free-wheeling, communications informal; documents would be distributed freely to members of the working group; and anyone with something to contribute could come to the party. As a historical account of the IETF noted, 'with this one document a swath was instantly cut through miles of red tape and pedantic process'.[7] Wikipedia similarly affirms a countercultural aversion to bureaucracy:

Wikipedia is not a moot court, and rules are not the purpose of the community. Instruction creep should be avoided. A perceived procedural error made in posting anything, such as an idea or nomination, is not grounds for invalidating that post. Follow the spirit, not the letter, of any rules, policies and guidelines if you feel they conflict. If the rules prevent you from improving the encyclopedia, you should ignore them. Disagreements should be resolved through consensus-based discussion, rather than through tightly sticking to rules and procedures. Wikipedia is not a bureaucracy.[8]

In reality, Wikipedia is clearly rules-based; it keeps written records of every possible transaction; and it is meritocratic not because of hacker-charismatic brilliance, but in the old-fashioned way: through the recognition of effort. All these traits correspond to the bureaucratic model. Until now sovereign authority has been defined as deriving from the will of the people, but a more accurate way of defining it would be to say that it represents *a combination of democracy and bureaucracy*. O'Mahony has noted that in order to introduce a bureaucratic basis of authority into a community form, members must design democratic mechanisms to limit that basis of authority.[9] In Debian, for example, the power of the project leader is limited in four ways: leaders must defer to the tribe; they have limited authority over technical matters; members can recall leaders; and the authority of leaders is counterbalanced by that of the Technical Committee.

Modern bureaucracy has had a complicated impact on the world. When large administrative entities ceased operating as part of feudal systems (as in imperial China), they organised qualified and impersonally selected officials according to standardised written records and rules, and promoted an ethic of responsibility and duty.[10] Not only was modern bureaucracy thus central to the expansion of industrial capitalism and Western imperialism: it also championed ideals such as meritocracy and egalitarianism.[11] Nonetheless, popular images of 'bureaucracy' conjure tropes of depersonalisation, centralisation, hierarchy, routine work, rigid procedures and lack of initiative. Bureaucracy has been derided by writers at opposite ends of the ideological spectrum: for Castoriadis it reproduces social inequality, whilst in Hayek's eyes it inhibits individual liberty.[12]

In the context of the contemporary autonomy imperative, bureaucracy was seen as antithetical to the non-hierarchical and individualistic 'new spirit of capitalism'.[13] Accordingly a range of alternatives issued from management research and the social sciences. Chief amongst these was that proposed by Manuel Castells, who invoked the development of a new organisational logic, the *networked enterprise*, which emerged in conjunction with informational capitalism. Networked enterprises comprised both new forms of individual involvement (flexible work) and new work practices (flexible production). According to this scenario, hierarchical organisations, under pressure from the combined forces of individualism, globalisation and the decentralising power of ICT, were no longer central institutions for the production of goods and services,.[14]

This was also to be the age of *post-bureaucratic organisations*, which were characterised by the search for consensus through dialogue (rather than through acquiescence to authority), an emphasis on organisational mission (rather than on rules and regulations), open and visible peer review processes (rather than hierarchical appraisal), open and permeable boundaries (rather than definite and impermeable boundaries). Post-bureaucratic organisations swam like fishes in the fast-paced, ever-changing 'knowledge economy'.[15] From management literature came homilies on decentralising authority, embracing participatory management, and empowering citizens by pushing control out of the bureaucracy into the community.[16] Economic, social, political and personal vitality would be best achieved by the generalisation of the enterprise form to all forms of conduct.[17] Such calls coincided with neoliberal attacks on the public sector in the form of deregulation and restructuring of public assets.

This raises the question of networks. In Castells's view, the networking logic induces a social determination of a higher level than that of the specific social interests expressed throughout the networks. Presence or absence in networks and the dynamics of each network vis-a-vis others are critical sources of domination and change in our society, 'a society that, therefore, we may properly call the network society, characterised by the preeminence of social morphology over social action'.[18] This applies not only to networked firms, but also to states and to social movements. Castells argues that networks represent the means by which social movement activists implement in the everyday the goal that is being sought out, by establishing spaces and practices that are themselves democratic and anti-hierarchical. Networks are potential forms for future democratic societies, in which people keep their autonomy, and conduct their debates and votes without the intermediation of professional politicians.[19] This book has shown that though anti-authoritarian groups challenge domination, authority is always present in autonomous communities. But the fact is, groups need leaders. What they don't need are *illegitimate* forms of authority, such as cliques. Barbara Epstein noted that the anti-leadership ideology of anti-globalisation activists could not eliminate leaders, but it could incite movements to deny that they have leaders, resulting in the undermining of democratic constraints on those who assume leadership roles, while 'also preventing the formation of vehicles for recruiting new leaders when the existing ones become too tired to continue'.[20] Organisational formats matter.

Networks may indicate which actors are central or strategically placed, but they cannot offer an understanding of how people justify this privileged position in relation to others. This can only be understood by examining the relationship between an organisational structure and the role a person occupies within it. What this boils down to is that assertions of the demise of the bureaucratic form are premature. In the first place, networked enterprises are still huge bureaucracies. Second, Yannis Kallinikos has argued that a network does not constitute an alternative form of *organisation*:

> unless a network is constituted as a unit of jurisdictional responsibility (which would mean that it would become some sort of bureaucracy), it is destined to remain no more than a social arrangement or practice, a strategy, as it were, for the reallocation of resources

in an informational economy.[21] Bureaucracy is a uniquely innovative organisational form and practice whose central characteristic, the separation of the organisational role from the concrete person, is not only still dominant, but highly flexible. The relationship between the individual and the organisation is based on *selectivity* (people are expected to suspend demands which do not pertain to their role and act following precise guidelines), *mobility* (a role can be transferred across organisations) and *reversibility* (jobs can be altered or withdrawn, and individuals can themselves leave).[22]

Bureaucracy is not always governed by high and invariable centralisation. There have been examples in the 1950s and 1960s of decentralisation, flexibility and idiosyncratic behaviour, which March and Olsen called 'organised anarchies'.[23] In the contemporary period, milder forms of domination such as 'soft bureaucracy' have been identified: organisations where 'processes of flexibility and decentralisation co-exist with more rigid constraints and structures of domination'.[24] In the end, then, the centralisation or distribution of authority, the more or less strong prevalence of routines and standardised behaviour, are important but derivative functions of the foundation of bureaucratic organisation: the clear separation of the individual from the organisation and the clear terms (selective, mobile and reversible) by which individuals are tied to organisations.[25]

Here lies the originality of online tribal bureaucracy: *the injection of charismatic authority into bureaucratic models re-establishes the connection between roles and persons.* This central point has been overlooked by the large numbers of organisation scholars who have conducted research into the governance of community forms of production. Where would Daily Kos and Wikipedia be without their charismatic founders? Would they exist at all? The authority of such personages is so great that their very appearance in a dispute inspires quasi-religious awe and quietens even the most argumentative disputants. Similarly, within the ultra-meritocratic Debian project, tribal elders occupy positions and command respect which go beyond a mere functional or role-based legitimacy. In other words, roles in tribal bureaucracies are selective (they follow guidelines), but not – in the case of charismatic founders and their lieutenants – mobile or reversible (the roles are unique and non-transferable).

These potentially autocratic characteristics are balanced by the introduction of collectivist or democratic elements into bureaucratic structures: the dislike of hierarchy combined with formal equality between participants. A third characteristic, permanent conflictuality, will be addressed later in this chapter. All these features clearly distinguish online tribal bureaucracies from post-bureaucratic organisations. The autonomy of workers in post-Fordist teams and the autonomy of consumers of social networking devices and platforms can only develop along avenues devised by senior management. The structures of resource allocation, control and decision making are hierarchical and centralised. In contrast, in online tribal bureaucracies, the enactment of policy requires individual decisions by community members to 'stick'. Even in cases in which a rule has been unilaterally decided by a central authority, it requires the assent of the governed to become policy, because enforcement is always highly decentralised.[26]

Costs and Benefits

This section outlines the costs and benefits of tribal bureaucracies, considered as productive units, in relation both to traditional bureaucracies or corporations and to traditional collectivist groups or communes. In general terms tribal bureaucracies combine elements of these other organisational formats. The most obvious difference between collectivist projects and corporate bureaucracies is that they are *volunteer* projects. *Participants want to participate.* The open structure of online projects enables them to match people with work relevant to their interest and expertise.[27] Online tribal bureaucracies are bare-bones operations: they are not hobbled by the same costs as corporate bureaucracies; but as a result they do not offer participants the same level of support as corporations do. Online projects are not legally responsible for the welfare of participants. Participants feel loyalty towards the health of the project, not that of other participants. In the same way, there is no need to retain underperforming individuals, or make sure they have something to do, as may be the case in public administrations. But it is also not easy to exclude underperformers from online tribal bureaucracies. The solution to poor performance in online projects, usually, is to encourage new people to participate, and to hope that the underperformers drift off.

The most obvious example of a cost borne by corporations derives from their hierarchical structure. *Employees in corporate bureaucracies do not own their authority.* They can only use it because it has been delegated to them by their boss and they themselves occupy the relevant position.[28] In contrast, when Debian developers fix a bug, when Wikipedia editors revert vandalism, when Daily Kos users uprate a diary and contribute towards putting it on the site's front page, they feel that they are making a uniquely useful contribution to the project. In other words, there is a much better alignment between the goals of participants and the overall goals of the project. Tribal bureaucracies are thus free from the problems which ail corporate bureaucracies, such as 'petty officials' inattentiveness to clients' demands ... loafing on the job, featherbedding, developing norms to limit production, stealing from a firm by employees, or appropriation of the firm's resources for personal use, excessive attention to rules rather than goals'.[29] This is somewhat offset by the problem of vandalism from *outsiders*, especially in projects to which anonymous users can contribute, such as Daily Kos and Wikipedia. Furthermore, there is little that leaders can do to oblige participants to perform unpopular tasks, as we saw in Debian with the question of training newcomers. Responsiveness is also not optimal, being very much a function of the individual: we may recall certain Debian developers' lack of response to requests. This is also a function of Debian's small size and barriers to entry to non-specialists. In mass projects that require no special initial expertise, such as Wikipedia, there is a high probability that any task, no matter how small, will find a volunteer.

Another type of corporate cost not borne by tribal bureaucracies can be called representational, in the sense that corporate actors like to be *perceived* as initiating or taking part in activities deemed to be dynamic or innovative. On the one hand this stems from the fact of being accountable to taxpayers and shareholders (which also results in corporations being submitted to audits, reviews and investigations, unlike online tribal bureaucracies), but it also derives from the internal dynamics of corporate bureaucracies, which pit managers against one another in competition for material resources, competent staff and promotions. Successful corporate leaders are members of action groups, working parties, committees and boards, and are enthusiastic proponents of the latest innovations in management science. While these activities may *look* productive, they are open to the charge that they mostly represent the adoption of business fads and short-term initiatives which bring no serious benefit to anyone, save the 'dynamic' corporate actors who take part in them. No such pressure exists in online tribal bureaucracies.

A final difference between corporate and tribal bureaucracies stems from the distribution of law enforcement. Traditionally a state Leviathan creates order by threatening to punish recalcitrant individuals. However, centralisation implies distance: there are limits to how much information can be gathered by a state, and even if all infractions are detected, Erikson and Parent point out that there is an 'endless complexity to wade through and adjudicate' because 'the limitless variation in malfeasance makes calibration difficult'.[30] The solution is to adopt a restricted repertoire of coarse penalties, so that authorities ignore the details of local circumstance. In online tribal bureaucracies, policing is much more precise, and personal. Phalanxes of mini-Leviathans with their finger on the ban-trigger are ready to enforce the law. Are these the birth pangs of a system of total control through mutual surveillance of each other's actions? Wikipedia fits that model, and it should be noted that distributed policing of norms and rules is being inculcated from an early age: virtual environments for children such as 'Club Penguin' also depend on a distributed caste of informants, known as special agents, who report norm violations to site owners, and whose special status provides them with added tools and access. However, while the online actions of contributors can be logged, their *physical movements* remain invisible to their co-workers, in marked contrast to private and public corporation employees. Not only are the communications of office workers monitored and recorded: workers are subjected to (and take part in) constant informal surveillance, by those in surrounding cubicles, of their social interactions and work patterns, as only middle- and upper-echelon managers are provided with enclosed offices. While the gains in efficiency of decreased physical monitoring in online projects are unclear, the 'privacy dividends' are obvious.

The major advantage held by online tribal bureaucracies over traditional collectivist groups, from which many others have flowed, is that *distributed networking has resolved the issue of absorbing increases in numbers of participants.* Traditionally, egalitarian organisations do not 'scale up' very well.[31] Lists, wikis and weblogs were expressly designed to enable the coordination of thousands of contributions. The bane of collectivist organisations – interminable meetings – is instantly overcome. Online communication does not have to occur synchronously, so not everyone has to be in the same room at the same time. People can take part in discussions whilst working on other parts of the project. Aside from improved time management, what other changes has digital communication wrought? Our understanding of the differences between online and offline collectivist groups will be helped by Rothschild-Whitt's characterisation of collectivist organisations.[32] Communes had highly selective *recruitment criteria* in order to ensure that new entrants shared the group's collectivist values. Recruitment was based on friendship and social–political values rather than specialised training, certification or universalised standards of competence. Online tribal bureaucracy recruitment procedures synthesise agreement with values and meritocracy; friendship also no doubt plays a part in supporting candidates. Candidates for Debian developership are examined for both their technical proficiency and their conformity to the spirit of the project. The conformity to project goals of Wikipedia candidates to adminships and other offices is assumed: what is vetted during the process is a candidate's past behaviour. In both cases, the process can take several months. Conversely, for non-administrative members of a community weblog or wiki, registering and participating is open and instant.

The notion that decision-making should ideally be consensual rendered conflict more threatening in collective organisations.[33] The prevalence of interpersonal tension in such groups has long been noted. Coleman suggests that communes are unstable because their structural elements are not *positions* but *persons*, and that 'while a position as an abstract entity is not subject to vagaries of sickness, temperament, or mental instability, persons are'.[34] Since online tribal bureaucracies combine communal and charismatic elements (personalities, especially of leaders, are important) and bureaucratic features, such as revocable positions, they are unstable, but not as emotionally fraught as communes.

In terms of the enforcement of standards of behaviour, norms such as 'netiquette' and 'wikiquette' are central to online projects, but the sheer size of online tribal bureaucracies renders the wholesale rejection of formal rules championed by collectivist groups unrealistic. As the size of a group increases, it becomes less likely that members will share a commonality of interest. A serious flaw of normative control is that social sanctions such as ostracism will have the greatest impact on people who value relations with other members. It is hard to punish a loner or transient effectively. In other words, norms assume a symmetry of interest within groups.[35] And since anonymity places less social constraint on action, it is easier to transgress a social boundary before strangers.[36] Online tribal bureaucracies need rules.

Finally, collectivist organisations rejected *stratification and dif-ferentiation*. Embracing egalitarianism meant there could be no specialisation of tasks, no experts and no hierarchy. Online tribal bureaucracies reject outside expertise, but embrace the home-grown variety. In terms of administrative power, rather than no one having authority over others, the distribution of enforcement means that everyone has some measure of authority over everyone else. This inevitably generates conflicts.

The Role of Conflict

It has long been accepted as a truism of Internet research that the anonymity made possible by the Net has a disinhibiting effect.

Internet users have the unusual capacity to release as much or as little of their true identity as desired, enabling them to safeguard themselves from the negative reactions they may encounter when these identity cues are salient in daily life. Because of the reduced visibility of cues such as class and gender, people are held to be less aware of their engagement in social interaction and less interested in achieving consensus.[37] For example, expressions of prejudice increase online, as people are less likely to exhibit discriminatory behaviour when their identity is knowable. The extended use of the Internet by extremist and hate groups can be attributed to anonymity.[38]

For project members, conflicts operate as a judicial sequence of action: a claim is made, the facts are examined, deliberation ensues. Eventually a resolution is reached and is accepted, producing compromise and appeasement, or is rejected, producing an escalation. For chiefs, conflicts serve as tests, allowing them, as Boltanski and Thévenot would say, to measure their *presumption to lead* against reality.[39] In bureaucracies, acquiring a leadership role means that people are elevated above the rest of the staff: their presumption to decide is vindicated by a recruitment committee. Online tribalism has an equalising effect and systematically questions presumptuousness. This makes it difficult to lead. On Daily Kos, the distribution of administrative power was highly prejudicial to dialogue between opponents. Weblogs in which disagreement can occur without vigilante cliques exterminating one another are those which have moderators to keep the peace, by ensuring that norms of fairness and politeness are systematically respected. It does not seem that this was the case on Daily Kos during the Democratic primary, possibly because of the size of the site. Debian leaders found it very hard to ban one of their own. Wikipedia resembles Daily Kos in that trusted participants are given access to tools that allow them to ban others. However, Wikipedia's charismatic founder is much less reluctant to use his personal authority to wade into battle and ban offenders than the founder of Daily Kos.

The online projects selected in this book to illustrate the dynamics of online authority are the largest and most prestigious in each of their respective fields. The *conflicts* evoked in the case studies were selected because each one condensed tensions operating at the heart of the project. Interestingly, all these defining events centred around a questioning of traditional gender roles. Alegre was a woman who refused to back down from a fight; SL was a male who expressed his innermost emotions; Durova was a woman intent on defending her sister admins from attack. These individuals were condemned by their respectives projects because of their perceived breaches of norms, such as failing to engage in deliberative dialogue (Alegre), failing to desist from spamming mailing lists (SL) or failing to give people the benefit of the doubt (Durova). But how did these individuals defend or justify these breaches? Essentially by arguing that they had no choice, as they were confronting sexism, cruelty, mindless vandalism, spite. In other words, the most violent conflicts in online projects pit the logic of the project against what individuals perceive to be the worst kind of injustice, that generated by *archaic force*.

But what of less momentous events? Management research divides conflicts in organisations into *task conflict* (having to do with work content), *affective conflict* (deriving from emotional relationships) and *process conflict* (concerning the approach to the task). It has been observed, for example, that when co-workers are friends there is more likelihood that affective conflict will occur. The distance separating members of distributed teams leads to more task and process conflicts because of 'different perspectives, inconsistent norms, incongruent temporal rhythms, reduced familiarity and demographic heterogeneity'.[40] In the remainder of this section, I examine how tribal bureaucracies are affected by task, affective and process conflict.

Task conflict in corporate bureaucracies can be positive when it obliges people to consider alternatives and consequences more fully, but it may degenerate into affective or process forms of conflict. Management research considers that distance and contextual differences will render it unmanageable in distributed teams.[41] Autonomous producers in online tribal bureaucracies can be very protective of their work, and expertise is often used as a weapon in content disputes. However, there are clear differences between instances where knowledge of specialised content is widely recognised as a positive attribute (see the primitivism.com archive and Debian) and others in which expertise is a much more controversial notion, such as on Daily Kos or Wikipedia. What remains constant throughout the examples we have examined is that the direction of a project is relatively uncontroversial. Though they may disagree on the details, all the participants agree on the finality of projects: producing free content. In this sense, online tribal bureaucracies are much akin to collectivist organisations, where ideological homogeneity meant that disputes were likely to be of the affectual or process variety.

In corporate settings, affective or emotional conflict is not encouraged. It is said to sap productivity, cause anxiety and hostility, and consume time and energy.[42] At the same time, members of distributed teams can easily avoid interacting with one another. This contrasts with online tribal bureaucracies. In theory, participants in online projects can work separately on their own sub-projects. However, since *strategy and orders do not originate from the summit of a corporate hierarchy, but are collectively debated*, such projects depend on the existence of electronic agora or meeting places; centralised sites such as blog comment threads, mailing lists or noticeboards, where everyone is meant to speak up, and where, in the grand old Usenet tradition, emotions run deep and flames run high. In short, affective conflict is rife in online tribal bureaucracies.

Process conflict in corporations can be defined as inefficiency resulting from confusion about resources and responsibilities.[43] Management researchers suggest that it may be worsened in corporate distributed teams by divergent perspectives and communication challenges, though shared team identity helps to bridge distance.[44] In reality, it is doubtful that in a hierarchical and centralised corporation there would be much scope for substantive disagreement about who is in charge, and how things should be done. This is exactly what does happen in online tribal bureaucracies. A recurrent theme of the conflicts we have examined is the betrayal of the projects' democratic promise. Two main reasons are advanced. First, a *clique* is said to be controlling the project, and stifling dissent. A second grievance is that procedures are not being consistently applied to all participants, either because of the existence of the suspected cabal, or because of the unpredictability of the charismatic founder.[45] These arguments go beyond the critique of a perceived lack of procedural predictability: they invoke a higher principle of equality or fairness, reminding us that online projects are not just organisational forms, but also political spaces. This issue is looked at in greater detail in the next section.

Conflict in online tribal bureaucracies can be destructive, as in communes, but it can also be channelled positively. Studies of Usenet flame wars have pointed out that conflict is not necessarily a bad thing, as it can lead to the resolution of disagreements, the establishment of consensus, the clarification of issues, and the strengthening of common values.[46] Flame wars can serve as rites of passage into the tribe for new members; the ease with which people can be excluded from online networks signifies that trolls, flamers, and sockpuppets can reinforce the unity of the project. The tribal constitution of an antagonistic other results in a paranoid view of the world: enemies (the state, Republicans, Ubuntu, Wiki-watchdog sites) are always out there, waiting and watching. This is particularly so in the case of weblogs, where enemies are actually used for the production of meaning when bloggers derisively comment on an opponent's views ('fisking'). The adversarial nature of weblogs renders them the most warrior-like of distributed online projects; not coincidentally, they also hold the most charismatic authority, and have the least restrictions on the expression of archaic force.

The Political Economy of Online Tribalism

Online tribal bureaucracies have emerged because of factors such as the spread of ICTs, globalisation, and disenchantment with bureaucratic authority. What is their likely wider social impact? Following Castells, Juris suggests that 'grassroots, network-based movements can be viewed as democratic laboratories, generating the political norms and forms most appropriate for the information age'.[47] The term 'laboratory' conveniently frees these 'norms and forms' from the strain of having an offline application. But what is being cooked up in the lab? Decentralised projects such as online tribal bureaucracies aim to give more control to individuals over political decision and economic production. In this book, I have equated these traits with 'tribalism'. Michel Maffesoli rightly saw that the feeling of proximity was a crucial aspect of neo-tribalism.[48] But he only defined these tribes as lifestyle aggregations, and did not perceive their potential as purposeful political entities.

In order to be proxemic, political processes need to be *deliberative*. Contemporary democracies suffer from a 'deliberative deficit', an absence of open spaces enabling the public to engage in open and critical discussion in order to exchange opinions and influence policy decisions.[49] The crisis of representative democracy generates interest in the kinds of close-knit discussions common in traditional tribes and villages. But deliberation also poses significant risks. First, deliberation enables the sequential updating of opinions. This raises the problem of *path-dependence*: the arbitrary nature of the order of speakers can influence the content and result of deliberation. Second, if the decisions of online self-governing authorities become an informal ongoing conversation instead of being manifested in discrete decisions such as final rules, statutes and judicial decisions, the quintessential legal provision of *due process* (being notified of the rule one is obligated to obey) becomes confused. The only way to be aware of the current status of a rule may be to continuously participate in decisions on it. And even if one participates regularly, there is no guarantee that the rule will be the same in a week.[50]

Thirdly, the *speed* of online deliberation may be problematic. Jaron Lanier remarks that online collectives have a jittery quality. Lanier considers that using a wiki for the writing of laws would be a terrifying prospect, as 'super-energised people would be struggling to shift the wording of the tax code on a frantic, never-ending basis'.[51] Lanier suggests that ordinary democracy may have a 'calming effect' on the political process, by reducing the potential for a collective to suddenly jump into an overexcited state in which too many rapid changes to answers end up cancelling each other out.

Finally, adopting deliberative procedures may have unexpected consequences for the popular appeal of collective decision. Nicolas Auray has pointed out that embracing tribal discussion poses the risk of weakening the sacred quality of a central political act: voting. Sacred things are protected by taboos and kept apart from collective deliberation. Intersubjectivity may be prohibited; people secretly deliberate alone in the booth, and this isolation endows voting with a sacred quality, separating it from previous moments of exchange and discussion. This ceremonial isolation and the verdict of chance also imparts a sense of *surprise* to political participation. In other words, for all their obvious qualities, deliberative procedures run the risk of making the political process less exciting, and hence less appealing.[52]

McLuhan's global village has been realised in online tribes in which everyone can talk to everyone else; but McLuhan had forgotten that villages are, above all else, the domain of mutual surveillance and gossip. This means that the quality of online deliberation is dependent on the widespread anonymity which is a central aspect of Internet activity.[53] Though anonymity makes it impossible to ascertain the sincerity of one's interlocutor, the absence of identity markers also signifies that the outcome of discussion does rest on the authority of the better argument. However, while those using the Internet must treat distant others on terms of equality, in practice the *responsibility* to do so is not immediately apparent: a discussant may simply opt to leave the debate, protected by anonymity.[54] If online tribal bureaucracies are to play a role greater than that of walled gardens filled with colourful avatars, and make significant contributions to democratic practice, they will have to re-examine the question of whether the benefits of anonymity are not outweighed by its disadvantages.

Apart from deliberation, tribalism has been defined in this book as the rejection of the market economy in favour of the cooperative (or 'peer') production of free goods. Cooperative production has induced writers in the Marxist tradition to describe the Internet in terms of an online struggle between cooperation, the 'essence' of human society, and competition; between information as public good and as commodity, between use and exchange value.[55] In the 1970s, members of collectivist groups were required to constantly shift gears by acting one way inside their collectives and another way outside. Their organisations were isolated examples of collectivism in an otherwise capitalist–bureaucratic context.[56] Online tribes are at the same time worse and better off in this respect. An essential condition for collectivist success was their integration into an interlocking network of cooperative organisations.[57] This objective has been realised on the Internet: there are ever more online peer-production ventures; there is ever more consciousness of the attractions of the free exchange of goods online. But online tribal projects are also completely integrated into the dominant economy, and the Internet ideology of freedom is the very embodiment of the new spirit of informational capitalism. Though zones of free exchange and deliberation exist, these are restricted to the sectors of online cultural or code production; the production of *hardware* is far from democratised. For the most part, oligarchs in liberal democracies and kleptocrats in dictatorships still control the means of production and communication, with the added benefit that volunteers are eagerly producing distributed content with no contracts, no pay and no benefits.

Online tribes represent an attempt to escape the clutches of the market and of corporate bureaucracy. It might be tempting to formulate some predictions as to the likely wider social impact of online tribes: they might remain separate enclaves, like the primitivists; or they might play the role of a public sphere feeding autonomous expertise into mainstream media politics, like the blogosphere; or they might offer an alternative to top-down decision-making processes, as is the case with Debian or Wikipedia. But these scenarios leave unaddressed the contradiction between autonomy and equality. For all the talk of flattening of hierarchies, it is unclear to what extent peer production challenges the class basis of industrial society. Clearly, not everyone has the financial and cultural resources to mix work and play and embrace a part-time artistic vocation as a Daily Kos contributing editor, Wikipedia administrator or Debian developer. In addition, the brute reality of network dynamics is that some nodes get all the attention. Neither illegitimate forms of power such as archaic force nor legitimate ones such as charismatic authority are democratic, as their source lies not in fair deliberation but in the reification of network dynamics or the importation of offline advantage. Ultimately they represent the surreptitious replacement of what are cultural forms of domination by so-called 'natural' or 'emergent' phenomena.

Finally, scenarios of the likely impact of online tribalism must take into account the great paradox of modernity: the critique of traditional figures of authority and the promotion of rationality, laity, and democratic public freedoms and social rights coincided with the development of the market.[58] Feudal society was holistic; capitalism introduced separation between people. The capitalist social model rejected domination based on transcendence situated outside human relationships, or on traditional forms of domination between people. Robert Castel argues that attempts to go beyond capitalism by emphasising closeness, mutual help, disinterested exchange and solidarity risk reverting to earlier pre-capitalist forms of domination.[59] The linking of roles to persons in online tribal bureaucracies could be interpreted as a regression to earlier modes of authority. A characteristic of pre-modern communities was that leaders practised patronage and the arbitrary exercise of rule.[60] Contrary to primitivist mythologising, the tribal world was one of mutual extermination.[61] What would happen if archaic forces such as patriarchal oppression were freed of the contractual limits imposed by markets?

So where does this leave us? With the need to reaffirm that autonomy does not only undermine equality. While autonomy can have that effect, the cases studied in this book show that wanting to cooperate with others in a self-directed way is a powerful human aspiration. In a world dominated by markets, more autonomy can therefore mean two things: on the one hand, more individualism, more inequality, more markets, with a few pockets of communal peer exchange on the fringes. On the other hand, a collective approach would encourage autonomous debate by tribal groups along lines not decided by bureaucracies or corporations. Such debate could examine alternatives to a market orthodoxy that holds that ever stronger economic expansion is the only way to live. Open discussion of the fact that the infinite creation of new needs is not sustainable, and of economic downscaling or 'a-growth' (as in atheism) would be enhanced. Ultimately the aim of such debate would be to formulate strategies to extend direct popular control over more aspects of existence, leading to the relocalisation or recommunalisation of everyday life. This might result in less luxury for some segments of humanity; but that may be the only way to create more autonomous and sustainable ways of living.

NOTES

Author's note: references given as '[online]' indicate that hyperlinks are available at www.plutobooks.com/cyberchiefs. All Web references were accessed on 18 September 2008.

Introduction

1. It should be noted that spontaneous online activity is only possible because the private companies which run the Net's series of tubes are not attempting to control online content – if they were, Net neutrality would be compromised, and so would autonomy on the Internet.
2. Some readers may wonder why no online gaming or virtual universes have been included in these case studies. The reasons are simple. Participants in these environments can freely create personas and objects, but they do so entirely at the discretion of these sites' owners. As a result, there can be no basis for meaningful self-direction of the site itself. This objection is, if possible, even more applicable in the case of 'social networking' platforms such as MySpace and Facebook. Such sites are not insignificant fads. On the contrary, they exemplify informational capitalism, as discussed in Chapter 1. But users of these services have very little autonomy. What recourse do banned users have? None. Can the terms of service – the laws of these worlds – be modified by users? Again, no. This may not be so different from the situation faced by someone who is banned from a weblog. But it is trivially easy to launch a new weblog. Launching a new social networking site is an altogether more difficult business.

Chapter 1

1. Manuel Castells, *The Power of Identity* (Vol. 2 of *The Information Age: Economy, Society and Culture*), 2nd edn, London: Blackwell, 2004, p. 370.
2. Ibid, p. 403.
3. Peter Dahlgren, 'Civic identity and Net activism', in Lincoln Dahlberg and Eugenia Siapera (eds), *Radical Democracy and the*

Internet: Interrogating Theory and Practice, New York: Palgrave Macmillan, 2007, pp. 55–72.

4. Manuel Castells and Araba Sey, 'From media politics to networked politics: the Internet and the political process', in Manuel Castells (ed.), *The Network Society: A Cross-Cultural Perspective*, Cheltenham: Edward Elgar, pp. 377–8.

5. Christian Fuchs, *Internet and Society: Social Theory in the Information Age*, London and New York: Routledge, 2008, p. 229.

6. Mikhail Bakunin, cited in Colin Ward, *Anarchy in Action*, London: Freedom Press, 1973, p. 22.

7. See Peter Marshall, *Demanding the Impossible: A History of Anarchism*, London: HarperCollins, 1992; Seán Sheehan, *Anarchism*, London: Reaktion Books, 2003.

8. William Whyte, *Making Mondràgon: The Growth and Dynamics of the Worker Cooperative Complex*, Ithaca, NY: Cornell University Press, 1992.

9. Guy Debord, *The Society of the Spectacle*, Detroit: Black and Red, 1983 [1967]; Raoul Vaneigem, *The Revolution of Everyday Life*, London: Left Bank Books and Rebel Press, 1983 [1967].

10. Steve Wright, *Storming Heaven: Class Composition and Struggle in Italian Autonomist Marxism*, London: Pluto Press, 2002.

11. Murray Bookchin, 'What is communalism?', *Green Perspectives*, No. 31 (1994), pp. 1–6.

12. Cornelius Castoriadis, *Figures of the Thinkable*, Palo Alto, CA: Stanford University Press, 2007.

13. George Katsiaficas, *The Subversion of Politics: European Autonomous Movements and the Decolonization of Everyday Life*, Atlantic Highlands, NJ: Humanities Press, 1997.

14. Amory Starr and Jason Adams, 'Anti-globalisation: the global fight for local autonomy', *New Political Science*, Vol. 25, No. 1 (2003), pp. 19–42.

15. Jerry Mander and Edward Goldsmith, *The Case Against the Global Economy: And for a Turn Toward the Local*, San Francisco: Sierra Club Books, 1996; Colin Hines, *Localisation: A Global Manifesto*, London: Earthscan, 2000.

16. Subcomandante Insurgente Marcos, 'Tomorrow begins today: closing remarks of the first International Encuentro for Humanity and Against Neoliberalism', in Juana Ponce de Leon (ed.), *Our Word Is Our Weapon: Selected Writings of Subcomandante Insurgente Marcos*, New York: Seven Stories Press, 2001, p. 112.

17. Harry Cleaver, 'The Zapatista effect: the Internet and the rise of an alternative political fabric', *Journal of International Affairs*, Vol. 5, No. 2 (1998), pp. 21–40.

18. Maria Garrido and Alexander Halavais, 'Mapping networks of support for the Zapatista movement', in Martha McCaughey and Michael D. Ayers (eds), *Cyberactivism: Online Activism in Theory and Practice*, New York and London: Routledge, 2003, pp. 164–84.

19. Castells, *Power of Identity*; Tom Mertes, *The Movement of Movements: A Reader*, London: Verso, 2004; Stevphen Shukaitis, 'Space. Imagination // Rupture: The cognitive architecture of utopian political thought in the Global Justice Movement', *University of Sussex Journal of Contemporary History*, No. 8 (2005) [online].

20. Castells, *Power of Identity*.

21. Manuel Castells, *The Internet Galaxy: Reflections on the Internet, Business and Society*, Oxford and New York: Oxford University Press, 2001; Pekka Himanen, *The Hacker Ethic: A Radical Approach to the Philosophy of Business*, New York: Random House, 2001. The countercultural celebration of small-scale technologies as instruments for the transformation of consciousness and community played a decisive role in shaping popular views of early computer networks. Fred Turner has shown how the ideals of the founders of the San Francisco Bay Area's *Whole Earth Catalog* percolated into the first 'alternative' online community, the Whole Earth 'Lectronic Link (WELL). See Fred Turner, *From Counterculture to Cyberculture: Stewart Brand, the Whole Earth Network, and the Rise of Digital Utopianism*, Chicago: University of Chicago Press, 2006.

22. Yochai Benkler, 'Coase's penguin, or, Linux and *The Nature of the Firm*', *Yale Law Review*, Vol. 112, No. 3 (2002), pp. 369–446.

23. Ibid, p. 444.

24. Ibid, p. 415.

25. Ibid, p. 378.

26. Ibid, p. 436.

27. Ibid, p. 435.

28. Cited in Glyn Moody, *Rebel Code: Linux and the Open Source Revolution*, London: Allen Lane, 2001, p. 28.

29. Dorothy Kidd, 'Indymedia.org: a new communications common', in McCaughey and Ayers (eds), *Cyberactivism*, pp. 47–69; Victor W. Pickard, 'United yet autonomous: Indymedia and the struggle to sustain a radical democratic network', *Media, Culture & Society*, Vol. 28, No. 3 (2006), pp. 315–36.

30. Barry Wellman, 'An electronic group is virtually a social network', in Sara B. Kiesler (ed.), *Culture of the Internet*, Mahwah, NJ: Lawrence Erlbaum, 1997, pp. 179–205.
31. Stanley Wasserman and Katherine Faust, *Social Network Analysis: Methods and Applications*, New York: Cambridge University Press, 1994.
32. Cited in Luc Boltanski and Eve Chiappello, *The New Spirit of Capitalism*, London: Verso, 2004 [1999], p. 3.
33. Electronic Frontier Foundation, *EFF News* #1.03 (7 March 1991) [online].
34. John Perry Barlow, *A Declaration of Independence of Cyberspace*, 1996 [online].
35. Nicholas Negroponte, 'Being digital: a book (p)review', *Wired*, February 1995, p. 182.
36. Sherry Turkle, *Life on the Screen: Identity in the Age of the Internet*, New York: Simon & Schuster, 1995; Mark Poster, 'Underdetermination', *New Media and Society*, Vol. 1, No. 1 (1999), p. 15.
37. Richard Barbrook, 'Cyber-communism: how the Americans are superseding capitalism in cyberspace', *Science as Culture*, Vol. 9, No. 1 (2000), pp. 5–40.
38. Richard Barbrook, 'The high-tech gift economy', *First Monday*, Vol. 3, No. 12 (1998) [online].
39. Richard Barbrook and Andy Cameron, 'The Californian ideology', in Peter Ludlow (ed.), *Crypto Anarchy, Cyberstates and Pirate Utopias*, Cambridge, MA: The MIT Press, 2001, pp. 363–87.
40. Mackenzie Wark, *A Hacker Manifesto*, Cambridge, MA: Harvard University Press, 2004.
41. Michael Hardt and Toni Negri, *Multitude*, New York: Penguin Books, 2004; Franco 'Bifo' Berardi, 'What is the meaning of autonomy today?', *Republicart* (September 2003) [online]; Toni Negri, *Negri on Negri: In Conversation with Ann Dufourmentelle*, New York and London: Routledge, 2004.
42. Michael Hardt and Toni Negri, *Empire*, Cambridge, MA: Harvard University Press, 2000, pp. 298–9.
43. Gilles Deleuze and Felix Guattari, *A Thousand Plateaus*, Minneapolis: University of Minnesota Press, 1987 [1980].
44. Sergio Bologna, 'The tribe of moles', in Sylvere Lotringer and Christian Marazzo (eds), *Italy: Autonomia – Post-Political Politics*, Brooklyn: Semiotext(e), 1980, pp. 36–61.
45. Hardt and Negri, *Empire*.
46. Ibid, 340.

47. Manuel Castells, *The Rise of the Network Society* (Vol. 1 of *The Information Age: Economy, Society and Culture*), 2nd edn, London: Blackwell, 2000.

48. Tiziana Terranova, 'Free labour: producing culture for the digital economy', *Social Text*, Vol. 18, No. 2 (2000), pp. 33–58.

49. Matthew Hindman, 'Open-source politics reconsidered', in Victor Mayer-Schoenberger and David Lazer (eds), *Governance and Information Technology: From Electronic Government to Information Government*, Cambridge, MA: MIT Press, 2007, pp. 183–207.

50. Terranova, 'Free labour', p. 36.

51. Fuchs, *Internet and Society*, p. 171.

52. Mirko Tobias Schafer and Patrick Kranzlmuller, 'RTFM! Teach-yourself culture in open source software projects', in Theo Hug (ed.), *Didactics of Microlearning*, Berlin and New York: Waxman, 2007, pp. 324–40.

53. Barlow, *Declaration of Independence*.

54. Dan Schiller, *Digital Capitalism: Networking the Global Market System*, Cambridge, MA: MIT Press, 1999.

55. Naomi Klein, *No Logo: Taking Aim at the Brand Bullies*, London: Flamingo, 2000; Basel Action Network, *Exporting Harm: The High-Tech Trashing of Asia*, 2002 [online].

56. Axel Bruns, *Gatewatching: Collaborative Online News Production*, New York: Peter Lang, 2005.

57. Boltanski and Chiappello, *New Spirit of Capitalism*.

58. Ibid.

59. Tom Hodgkinson, 'With friends like these', *Guardian*, 14 January 2008.

60. Henry Jenkins, 'Hustling 2.0: Soulja Boy and the crank dat phenomenom', *Confessions of an Aca-Fan*, 31 October 2007 [online].

61. François Dubet, *Les Inégalités Multipliées*, La Tour d'Aigues: Editions de l'Aube, 2001.

62. Barry Wellman, 'Physical place and cyber place: the rise of personalised networking', *Journal of Urban and Regional Research*, No. 25 (June 2001), pp. 227–52.

63. Castells, *Internet Galaxy*, p. 131.

64. Jan Fernback, 'Beyond the diluted community concept: a symbolic interactionist perspective on online social relations', *New Media & Society*, Vol. 9, No. 1 (2007), p. 54.

65. Howard Rheingold, *The Virtual Community: Homesteading on the Electronic Frontier*, Reading, MA: Addison-Wesley, 1993.

66. See Nancy Baym, 'The emergence of on-line community', in Steve Jones (ed.), *Cybersociety 2.0: Revisiting Computer-Mediated*

Communication and Community, Thousand Oaks, CA: Sage, 1998, pp. 35–68; Judith S. Donath, 'Identity and deception in the virtual community', in Marc Smith and Peter Kollock (eds), *Communities in Cyberspace*, New York: Routledge, 1999, pp. 29–59.

67. Darin Barney, *The Network Society*, Cambridge: Polity, 2004, p. 159.

68. Fernback, 'Beyond the diluted community concept'.

69. Etienne Wenger, *Communities of Practice: Learning, Meaning and Identity*, Cambridge: Cambridge University Press, 1998.

70. Jean Lave and Etienne Wenger, *Situated Learning: Legitimate Peripheral Participation*, Cambridge: Cambridge University Press, 1991.

71. Peter M. Haas, 'Introduction: epistemic communities and international policy coordination', *International Organization*, Vol. 46, No. 10 (1992), pp. 1–35.

72. Kevin Hetherington, 'Stonehenge and its festivals: spaces of consumption', in Rob Shields (ed.), *Lifestyle Shopping: The Subject of Consumption*, London: Routledge, 1992, pp. 83–98; Rob Shields, 'The individual, consumption cultures and the fate of community', in Shields (ed.), *Lifestyle Shopping*, pp. 99–113; Andy Bennett, 'Subcultures or neo-tribes? Rethinking the relationship between youth, style and musical style', *Sociology*, Vol. 33, No. 3 (1999), pp. 599–617.

73. Michel Maffesoli, *The Time of the Tribes: The Decline of Individualism in Mass Societies*, London, Thousand Oaks, CA and New Delhi: Sage, 1996.

74. Brian Holmes, *Unleashing the Collective Phantoms: Essays in Reverse Imagineering*, Brooklyn, NY: Autonomedia, 2007.

75. Marshal McLuhan, *Understanding Media: The Extensions of Man*, New York: New American Library, 1964.

76. Tyrone L. Adams and Stephen A. Smith, *Electronic Tribes*, Austin: University of Texas Press, 2008.

77. William H. Shaw, 'Marx and Morgan', *History and Theory*, Vol. 23, No. 2 (May 1984), pp. 215–28.

78. Pierre Clastres, *Society Against the State: Essays in Political Anthropology*, New York: Zone Books, 1987 [1974].

79. Public goods are non-excludable (anyone can benefit from them regardless of their contribution) and non-rival (use or consumption of the good does not reduce the amount available to others).

80. Yochai Benkler and Helen Nissenbaum, 'Commons-based peer production and virtue', *Journal of Political Philosophy*, Vol. 14, No. 4, 2006, p. 408.

81. Didier Demazière, Francois Horn and Marc Zune, 'Dynamique de développement des communautés du logiciel libre: Condition d'émergence et régulation des tensions', *Terminal*, Nos. 97–8 (October 2006), pp. 71–84.

Chapter 2

1. David Beetham, *The Legitimation of Power*, London: Macmillan, 1991.
2. Max Horkheimer, *Critical Theory: Selected Essays*, New York: Herder & Herder, 1972 [1936].
3. Erich Fromm, *Escape From Freedom*, New York: Henry Holt, 1941.
4. Theodor Adorno, Else Frenke-Brunswil, Daniel J. Levinson and R. Nevitt Sanford, *The Authoritarian Personality*, New York: Harper & Row, 1950, p. 971.
5. Richard Christie and Marie Jahoda, *Studies in the Scope and Method of the 'Authoritarian Personality'*, Glenco, IL: Free Press, 1954; Robert Altemeyer, *Right-Wing Authoritarianism*, Winnipeg, Canada: University of Manitoba Press, 1981.
6. Michel Crozier, Samuel Huntington and Joji Watanuki, *The Crisis of Democracy*, New York: New York University Press, 1975.
7. Alexander Mitscherlich, *Society Without the Father: A Contribution to Social Psychology*, New York: Harcourt, Brace, & World, 1969; Christopher Lasch, *The Culture of Narcissism: American Life in an Age of Diminishing Expectations*, New York: Norton, 1978.
8. Michael Hardt and Toni Negri, *Multitude*, New York: Penguin Books, 2004, p. 210.
9. Hans Magnus Enzensberger, 'Constituents of a theory of the media', *New Left Review*, No. 64 (1970), pp. 13–36.
10. Ibid.
11. Ibid.
12. Benjamin Barber, 'Which technology and which democracy?', Democracy and Digital Media Conference, MIT, 8–9 May 1998 [online].
13. Yochai Benkler, *The Wealth of Networks: How Social Production Transforms Markets and Freedom*, Yale: Yale University Press, 2006, p. 212.
14. Darin Barney, *The Network Society*, Cambridge: Polity, 2004.
15. Rick Levine, Christopher Locke, Doc Searls and David Weinberger, *The Cluetrain Manifesto: The End of Business as Usual*, Cambridge, MA: Perseus Books, 2000, p. 8.

16. Geert Lovink, 'The principles of notworking', Inaugural speech, Hogeschool van Amsterdam, 2005 [online].
17. James E. Rauch and Gary G. Hamilton, 'Networks and markets: concepts for bridging disciplines', in James E. Rauch and Alessandra Casella (eds), *Networks and Markets*, New York: Russell Sage Foundation, 2001, pp. 1–29.
18. John Dryzek, 'Legitimacy and economy in deliberative democracy', *Political Theory*, Vol. 29, No. 5 (2001), p. 664.
19. Manuel Castells, *The Power of Identity* (Vol. 2 of *The Information Age: Economy, Society and Culture*), 2nd edn, London: Blackwell, 2004, p. 156.
20. Luc Boltanski and Eve Chiappello, *The New Spirit of Capitalism*, London: Verso, 2004 [1999], p. 130.
21. Lawrence Lessig, *Code and Other Laws of Cyberspace*, New York: Basic Books, 1999, p.130.
22. Alexander Galloway, *Protocol: How Control Exists After Decentralisation*, Cambridge, MA: MIT Press, 2004.
23. Paul Baran, *On Distributed Communication*, Santa Monica, CA: Rand, 1964.
24. Galloway, *Protocol*, p. 38.
25. Max Weber, *Economy and Society: An Outline of Interpretive Sociology*, Berkeley, Los Angeles and London: University of California Press, 1978 [1922], p. 244.
26. Ibid, p. 241.
27. Ibid, p. 1121.
28. Ibid, pp. 262–6.
29. Paul M. Harrison, 'Weber's categories of authority and voluntary associations', *American Sociological Review*, Vol. 25, No. 2 (April 1960), pp. 232–7.
30. Weber, *Economy and Society*, p. 24.
31. James Coleman, 'Social inventions', *Social Forces*, No. 49 (1970), pp. 163–73.
32. Max Weber, 'The sociology of charismatic authority', in Hans Gerth and C. Wright Mills (eds), *From Max Weber: Essays in Sociology*, New York: Oxford University Press, 1946, p. 248.
33. James V. Downton, *Rebel Leadership: Commitment and Charisma in the Revolutionary Process*, New York: Free Press, 1974.
34. Stevphen Shukaitis, 'Space. Imagination // Rupture: the cognitive architecture of utopian political thought in the Global Justice Movement', *University of Sussex Journal of Contemporary History*, No. 8, 2005 [online].
35. Ibid.

36. Bernard Bass, *Leadership and Performance Beyond Expectations*, New York: Free Press, 1985, p. 242; Jay A. Conger and Rabindra N. Kanungo, *Charismatic Leadership in Organisations*, Thousand Oaks, CA: Sage, 1998.

37. Philip Smith, 'Culture and charisma: outline of a theory', *Acta Sociologica*, Vol. 43, No. 2 (2000), pp. 101–11.

38. Edward A. Tiryakian, 'Collective effervescence, social change and charisma: Durkheim, Weber and 1989', *International Sociology*, Vol. 10, No. 3 (1995), pp. 269–81.

39. William H. Swatos, 'The disenchantment of charisma: on revolution in a rationalised world', in Ronald M. Glassman and William H. Swatos (eds), *Charisma, History and Social Structure*, New York: Greenwood Press, 1986, pp. 129–46.

40. Boltanski and Chiappello, *New Spirit of Capitalism*.

41. Ronald Glassman, 'Legitimacy and manufactured charisma', *Social Research*, Vol. 42, No. 4 (1975), pp. 615–36; Joseph Bensman and Michael Givant, 'Charisma and modernity: the use and abuse of a concept', *Social Research*, Vol. 42, No. 4 (1975), pp. 570–614.

42. Ronald M. Glassman and William H. Swatos, 'Introduction', in Glassman and Swatos (eds), *Charisma, History and Social Structure*, p. 6.

43. Weber, *Economy and Society*, p. 241.

44. Bruce M. Kogut and Anca Metiu, 'Open-source software development and distributed innovation', *Oxford Review of Economic Policy*, Vol. 17, No. 2 (May 2001), pp. 248–64.

45. Richard S. Bell, 'Charisma and illegitimate authority', in Glassman and Swatos (eds), *Charisma, History and Social Structure*, p. 59.

46. Janet Abbate, *Inventing the Internet*, Cambridge, MA: MIT Press, 1999, p. 127.

47. William A. Hagstrom, *The Scientific Community*, New York: Basic Books, 1965.

48. Bernard H. Gustin, 'Charisma, recognition, and the motivation of scientists', *American Journal of Sociology*, Vol. 78, No. 5 (1973), pp. 1119–34.

49. Paul Hoffman and Susan Harris, 'The Tao of the IETF: a novice's guide to the Internet Engineering Task Force', RFC 4677, FYI 17, September 2006 [online].

50. Manuel Castells, *The Internet Galaxy: Reflections on the Internet, Business and Society*, Oxford and New York: Oxford University Press, 2001, p. 46.

51. Steven Levy, *Hackers: Heroes of the Computer Revolution*, Garden City, NY: Doubleday, 1984, p. ix.

52. Castells, *Internet Galaxy*, p. 40.

53. Ibid, p. 31.
54. Hoffman and Harris, 'The Tao of the IETF'.
55. Glyn Moody, *Rebel Code: Linux and the Open Source Revolution*, London: Allen Lane, 2001; Castells, *Internet Galaxy*, p. 48.
56. Hoffman and Harris, 'The Tao of the IETF'.
57. Free Software Foundation, 'The Free Software Definition' [online].
58. Moody, *Rebel Code*.
59. Smith, 'Culture and charisma'.
60. Zygmunt Bauman, *Intimations of Postmodernity*, London: Routledge, 1992.
61. Anthony P. Cohen, *The Symbolic Construction of Community*, London and New York: Tavistock, 1985.
62. Fredrik Barth, 'Introduction', in Fredrik Barth (ed.), *Ethnic Groups and Boundaries: The Social Organisation of Culture Difference*, Long Grove, IL: Waveland Press, 1969, pp. 9–38.
63. Yochai Benkler and Helen Nissenbaum, 'Commons-based peer production and virtue', *Journal of Political Philosophy*, Vol. 14, No. 4 (2006), p. 400.
64. Michael A. Froomkin, 'Wrong turn in cyberspace: using ICANN to route around the APA and the constitution', *Duke Law Journal*, Vol. 50 (2000), p. 17.
65. Vinton Cerf, quoted in John King, Rebecca E. Grinter and Jeanne M. Pickering, 'The rise and fall of netville', in Sara Kiesler (ed.), *Culture of the Internet: Research Milestones from the Social Sciences*, Mahwah, NJ: Lawrence Erlbaum, p. 16.
66. Hoffman and Harris, 'The Tao of the IETF'.
67. Bryan Pfaffenberger, '"If I want it, it's OK": Usenet and the (outer) limits of free speech', *The Information Society*, No. 12, pp. 365–86.
68. Ibid, p. 368.
69. Glyn Moody points out that many have a different image of Stallman: not so much playful mock saint as implacable Old Testament prophet, 'a kind of geek Moses bearing the GNU GPL commandments and trying to drag his hacker tribe to the promised land of freedom whether they want to go or not. His long hair, which falls thickly to his shoulders, full beard, and intense gaze doubtless contribute to the effect.' Moody, *Rebel Code*, p. 29.
70. Eric Raymond, *The Cathedral and the Bazaar: Musings on Open Source and Linux by an Accidental Revolutionary*, Sebastopol, CA, O'Reilly, 1999.
71. Michel Gensollen, 'L'économie des communautés médiatées', Rapport Final, CNRS Programme Interdisciplinaire 'Société de l'Information', April 2005.

72. Josh Lerner and Jean Tirole, 'The economics of technology sharing: open source and beyond', NBER Working Paper No. 10956, December 2004, p. 11.
73. Raymond, *The Cathedral and the Bazaar*.
74. Eric Raymond, cited in William C. Taylor, 'Inspired by work', *FastCompany*, No. 200 (October 1999), p. 200.
75. Juan Mateos-Garcia and W. Edward Steinmueller, 'Applying the open source development model to knowledge work', *SPRU Electronic Working Paper Series*, No. 94, University of Sussex, 2003.
76. Felix Stalder and Jesse Hirsh, 'Open source intelligence', *First Monday*, Vol. 7, No. 6 (2002) [online].
77. Raymond, *The Cathedral and the Bazaar*, p. 30.
78. Mateos-Garcia and Steinmueller, 'Applying the open source development model'.
79. Juan Mateos-Garcia and W. Edward Steinmueller, 'The open source way of working: a new paradigm for division of labour in software development', *SPRU Electronic Working Paper Series*, No. 92, University of Sussex, 2003.
80. Lessig, *Code and Other Laws*.
81. Eric Raymond, quoted in Moody, *Rebel Code*, p. 177.
82. Mateos-Garcia and Steinmueller, 'Applying the open source development model'.
83. Kasper Edwards, 'When beggars become choosers', *First Monday*, Vol. 5, No. 10 (2000) [online].
84. Raymond, quoted in Taylor, 'Inspired by work'.
85. Stalder and Hirsh, 'Open source intelligence'.
86. Didier Demazière, Francois Horn and Marc Zune, 'Dynamique de développement des communautés du logiciel libre: Condition d'émergence et régulation des tensions', *Terminal*, No. 97–8 (October 2006), p. 80.
87. Tim Berners-Lee, *Weaving the Web*, New York: HarperCollins, 1999; Lev Manovich, *The Language of New Media*, Cambridge, MA: MIT Press, 2001; Benkler, *Wealth of Networks*.
88. Quoted in Galloway, *Protocol*, p. 23.
89. See Pfaffenberger, '"If I want it, it's OK"'.
90. Popular contemporary MUDs include Ostromud, Discworldmud and Xyllomer.
91. Pavel Curtis, 'Mudding: social phenomena in text-based virtual realities', in *Proceedings of the Conference on Directions and Implications of Advanced Computing*, Berkeley, 5–7 May 1992.
92. Elizabeth S. Reid, 'Hierarchy and power: social control in cyberspace', in Marc Smith and Peter Kollock (eds), *Communities in Cyberspace*, London: Routledge, 1999, pp. 107–33.

93. Peter Kollock and Marc Smith, 'Introduction: communities in cyberspace', in Smith and Kollock, *Communities in Cyberspace*, pp. 3–28.
94. Bernard Conein, 'Relations de conseil et expertise collective: comment les experts choisissent-ils leurs destinataires dans les listes de discussions?', *Recherches Sociologiques*, Vol. 35, No. 3 (2004), pp. 61–74.
95. Benkler, *Wealth of Networks*, p. 228.
96. Ibid, p. 168.
97. Jaron Lanier, 'Digital Maoism: the hazards of the new online collectivism', *The Edge*, May 2006 [online].
98. David Beer and Roger Burrows, 'Sociology and, of and in Web 2.0: some initial considerations', *Sociological Research Online*, Vol. 12, No. 5 (2007), 2.14 [online].
99. David Shay and Trevor Pinch, 'Six degrees of reputation: the use and abuse of online review and recommendation systems', *ScTS Working Paper*, Cornell University, 2005.
100. Ibid, p. 19.
101. Ibid, p. 6.
102. Paul Resnick, Richard Zeckhauser, Eric Friedman and Ko Kuwabara, 'Reputation systems', *Communications of the ACM*, Vol. 43, No. 12 (December 2000), pp. 45–8.
103. Yochai Benkler, 'Coase's penguin, or, Linux and *The Nature of the Firm*', *Yale Law Review*, Vol. 112, No. 3 (2002), p. 394.
104. Lanier, 'Digital Maoism'.
105. Sergey Brin and Lawrence Page, 'Anatomy of a large-Scale hypertextual Web search engine', in *Proceedings of the Seventh International World Wide Web Conference*, 1998. Benkler calls this the 'peer production of ranking'. See Benkler, 'Coase's Penguin', p. 392.
106. Jon M. Kleinberg, 'Authoritative sources in a hyperlinked environment', *Journal of the ACM*, Vol. 46, No. 5 (1999), pp. 604–32.
107. Matthew Hindman, Kostas Tsioutsiouliklis and Judy A. Johnson, '"Googlearchy": How a Few Heavily-linked Sites Dominate Politics on the Web', mimeograph, Princeton University, 2003.
108. Kleinberg, 'Authoritative Sources', p. 606.
109. Robert Ackland and Mathieu O'Neil, 'Online collective identity: the case of the environmental movement', Australian National University, ADSRI Working Paper No. 4, 2008.
110. Bruno Latour, *Science in Action: How to Follow Scientists and Engineers through Society*, Cambridge, MA: Harvard University Press, 1987.

111. Daniel Brass and David Krackhardt, 'The social capital of twenty-first century leaders', in James G. Hunt, George E. Dodge and Leonard Wong (eds), *Out-of-the Box Leadership: Transforming the Twenty-First Century Army and other Top Performing Organisations*, Amsterdam: Elsevier, 1999, pp. 179–94.

112. Ronald S. Burt, *Structural Holes: The Social Structure of Competition*, Cambridge, MA: Harvard University Press, 1992.

113. Mark Poster, 'CyberDemocracy: the Internet and the public sphere', in David Holmes (ed.), *Virtual Politics: Identity and Community in Cyberspace*, London: Sage, 1997, pp. 212–29; see also Michael A. Froomkin, 'Habermas@discourse.net: towards a critical theory of cyberspace', *Harvard Law Review*, Vol. 116, No. 3 (2003), pp. 749–873; for a more critical assessment, see Lincoln Dahlberg, 'Computer-mediated communication and the public sphere: a critical analysis', *Journal of Computer-Mediated Communication*, Vol. 7, No. 1 (October 2001).

114. Jürgen Habermas, *The Theory of Communicative Action: The Critique of Functionalist Reason*, Cambridge: Polity Press, 1987, p. 145.

Chapter 3

1. Herbert A. Simon, 'On a class of skew distribution functions', *Biometrika*, Vol. 42, Nos. 3–4 (December 1955), pp. 425–40; Albert-Laszlo Barabàsi, *Linked: The New Science of Networks*, Cambridge, MA: Perseus, 2002.

2. Fredrik Liljeros, Christofer R. Edling, Luís A. Nunes Amaral, H. Eugene Stanley and Yvonne Åberg, 'The Web of Sexual Contacts', *Nature*, No. 411 (June 2001), pp. 907–8.

3. Barabàsi, *Linked*.

4. See Andrei Broder, Ravi Kumar, Farzin Maghoul, Prabhakar Raghavan, Sridhar Rajagopalan, Rayme Stata, Andrew Tomkins and Janet Wiener, 'Graph structure in the Web', in *Proceedings of the Ninth International World Wide Web Conference*, Amsterdam: Elsevier, pp. 309–20; Bernardo A. Huberman, 'Scale free networks: structure, dynamics and search', HP Labs, 2002; Matthew Hindman, Kostas Tsioutsiouliklis and Judy A. Johnson, '"Googlearchy": how a few heavily-linked sites dominate politics on the Web', Annual Meeting of the Midwest Political Science Association, Chicago, IL, 4–6 April 2003; Clay Shirky, 'Power laws, weblogs, and inequality', in Jodi Dean, Jon W. Anderson and Geert Lovink (eds), *Reformatting Politics: Information Technology and Global Civil Society*, New York and London:

Routledge, 2006; Matthew Hindman, '"Open-source politics" reconsidered', in Victor Mayer-Schoenberger and David Lazer (eds), *Governance and Information Technology: From Electronic Government to Information Government*, Cambridge, MA: MIT Press, pp. 183–207.

5. Barabàsi, *Linked*.
6. Shirky, 'Power laws, weblogs, and inequality', p. 36.
7. David M. Pennock, Gary W. Flake, Steve Lawrence, Eric J. Glover and C. Lee Giles, 'Winners don't take all: characterizing the competition for links on the web', *Proceedings of the National Academy of Sciences*, Vol. 99, No. 8, pp. 5207–11.
8. Barabàsi, *Linked*.
9. Mustafa Emirbayer and Jeff Goodwin, 'Network analysis, culture, and the problem of agency', *American Journal of Sociology*, No. 99 (1994), pp. 1411–54; Wouter De Nooy, 'Fields and networks: correspondence analysis and social network analysis in the framework of field theory', *Poetics*, Vol. 31, Nos. 5–6 (2003), pp. 305–27.
10. Huberman, 'Scale free networks', p. 5.
11. Albert-Laszlo Barabàsi and Eric Bonabeau, 'Scale-free networks', *Scientific American*, May 2003, p. 54.
12. Ibid, p. 54.
13. Shirky, 'Power laws, weblogs, and inequality', p. 35.
14. Barabasi, *Linked*, p. 245.
15. Felix Stalder and Jesse Hirsh, 'Open source intelligence', *First Monday*, Vol. 7, No. 6 (2002) [online].
16. Glyn Moody, *Rebel Code: Linux and the Open Source Revolution*, London: Allen Lane, 2001, p. 175.
17. Eric Raymond, *The Cathedral and the Bazaar: Musings on Open Source and Linux by an Accidental Revolutionary*, Sebastopol, CA: O'Reilly, 1999.
18. Peter Muhlberger, 'Human agency and the revitalisation of the public sphere', *Political Communication*, Vol. 22, No. 2 (2005), pp. 163–78.
19. Ibid, p. 167.
20. Lucas D. Introna and Helen Nissenbaum, 'Shaping the Web: why the politics of search engines matter', *The Information Society*, Vol. 16, No. 3 (2000), p. 169–85.
21. Ibid, p. 171.
22. Sergey Brin and Lawrence Page, 'Anatomy of a large-scale hypertextual Web search engine', in *Proceedings of the Seventh International World Wide Web Conference*, 1998, pp. 107–17.
23. Google, 'Corporate information: technology overview' [online].
24. Hindman et al., '"Googlearchy"'.

25. Robert K. Merton, 'The Matthew effect in science', *Science*, Vol. 159, No. 3810 (January 1968), pp. 56–63.
26. Hindman, '"Open-source politics" reconsidered'.
27. Santo Fortunato, Alessandro Flammini, Filippo Menczer and Alessandro Vespignani, 'The egalitarian effect of search engines', Technical Report, arXiv.org e-Print Archive, 2005 [online].
28. Yochai Benkler, *The Wealth of Networks: How Social Production Transforms Markets and Freedom*, Yale: Yale University Press, 2006, p. 241.
29. Ibid, p. 224.
30. Ibid, p. 237.
31. Ibid, pp. 238–9.
32. Ibid, p. 246.
33. Henry Farrell and Daniel W. Drezner, 'The power and politics of blogs', *Public Choice*, Vol. 134, Nos. 1–2 (2008), pp. 15–30.
34. Benkler, *Wealth of Networks*, p. 238.
35. Ibid, p. 239.
36. Shirky, 'Power laws, weblogs, and inequality', p. 40.
37. Ibid.
38. Emirbayer and Goodwin, 'Network analysis', p. 1428.
39. Shirky, 'Power laws, weblogs, and inequality', p. 40.
40. Emirbayer and Goodwin, 'Network analysis', p. 1425.
41. Alexander Galloway, *Protocol: How Control Exists After Decentralisation*, Cambridge, MA: MIT Press, 2004, p. 122.
42. Karim R. Lakhani and Robert G. Wolf, 'Why hackers do what they do: understanding effort and motivation in free/open-source software projects', in Joseph Feller, Brian Fitzgerald, Scott A. Hissam and Karim R. Lakhani (eds), *Perspectives on Free and Open Source Software*, Cambridge, MA: MIT Press, 2005, pp. 3–21.
43. Hindman, '"Open-source politics" reconsidered', p. 199.
44. Offline deliberation is similarly far from democratic: the more intensive the form of participation, the greater the tendency to over-represent high-status members of the population. See Jack A. Nagel, *Participation*, New York: Prentice Hall, 1987. A survey of 15,000 North Americans found that of all forms of political participation, voting is most equally distributed among social classes. After campaign contributions, the most deliberative form of participation, 'membership on a local board', was most clearly linked to high income and status. See Sydney Verba, Kay Lehman Scholzman and Henry E. Brady, *Voice and Equality: Civic Voluntarism in American Politics*, Cambridge MA: Harvard University Press, 1995.

45. David Beetham, *The Legitimation of Power*, London: Macmillan, p. 105; Pierre Bourdieu, *Outline of a Theory of Practice*, Cambridge: Polity Press, 1977, p. 164.

46. Pierre Bourdieu, *The State Nobility: Elite Schools in the Field of Power*, Stanford: Stanford University Press, 1996.

47. Nancy Fraser, 'Rethinking the public sphere: a contribution to the critique of actually existing democracy', in Craig Calhoun (ed.), *Habermas and the Public Sphere*, Cambridge, MA: MIT Press, 1992, p. 116. Radical and/or feminist political theorists have questioned the disciplining implicit in the designation of a *particular* form of communication as *the* rational and democratically legitimate norm; in order to be equally included, some participants must be more disciplined than others so as to fit into an idealised deliberative mode, resulting in the exclusion or repression of voices deemed illegitimate – that is to say irrational, non-democratic, private. See Lincoln Dahlberg, 'The Internet, deliberative democracy and power: radicalising the public sphere', *International Journal of Media and Cultural Politics*, Vol. 3, No. 1 (2007), pp. 47–64.

48. Geoff Eley, 'Nations, publics, and political cultures: placing Habermas in the nineteenth century', in Calhoun (ed.), *Habermas and the Public Sphere*, p. 115.

49. Craig Calhoun, 'Habitus, field and capital', in Craig Calhoun, Edward LiPuma and Moishe Postone (eds), *Bourdieu: Critical Perspectives*, Chicago: University of Chicago Press, 1993, p. 63.

50. Albert Ogien, 'Une critique sans institution et sans histoire?', in Jean Lojkine (ed.), *Les Sociologies critiques du capitalisme*, Paris: Presses Universitaires de France, 2002, pp. 161–75.

51. Loïc Wacquant, 'De l'idéologie à la violence symbolique: culture, classe et conscience chez Marx et Bourdieu', in Lojkine (ed.), *Les Sociologies critiques du capitalisme*, pp. 25–40.

52. See Pierre Bourdieu, 'The field of cultural production, or: the economic world reversed', *Poetics*, Vol. 12, Nos. 4–5 (1983), pp. 311–56; Pierre Bourdieu, 'The social space and the genesis of groups', *Theory and Society*, Vol. 14, No. 6 (1985), pp. 723–44.

53. Victor Turner, *Dramas, Fields and Metaphors*, Ithaca, NY: Cornell University Press, 1974, p. 135.

54. Bourdieu, 'The social space and the genesis of groups'.

55. Pierre Bourdieu, 'Social space and symbolic power', *Sociological Theory*, Vol. 7, No. 1 (Spring 1989), p. 23.

56. Corinna Di Gennaro and William Dutton, 'The Internet and the public: online and offline political participation in the United Kingdom', *Parliamentary Affairs*, Vol. 59, No. 2 (2006), pp. 299–313.

57. Bernard Lahire, *La Culture des individus, dissonances culturelles et distinction de soi*, Paris: Editions La Découverte, 2004.

58. Beverley Skeggs, *Class, Self, Culture*, London: Routledge, 2004.

59. Ibid, p. 131.

60. Beverley Skeggs, 'Exchange, value and affect: Bourdieu and "the self"', in Lisa Adkins and Beverley Skeggs (eds), *Feminism After Bourdieu*, Oxford: Blackwell, 2004, pp. 75–97.

61. Ibid.

62. Beverley Skeggs, 'Uneasy alignments, resourcing respectable subjectivity', *GLQ: A Journal of Lesbian and Gay Studies*, Vol. 10, No. 2 (2004), pp. 291–98. See also Chris Haylett, 'Illegitimate subjects? Abject whites, neoliberal modernisation, and middle-class multiculturalism', *Environment and Planning D: Society and Space*, No. 19 (2001), pp. 351–70; Matt Wray and Annalee Newitz, *White Trash: Race and Class in America*, London: Routledge, 1997.

63. Michael A. Froomkin, 'Habermas@discourse.net: towards a critical theory of cyberspace', *Harvard Law Review*, Vol. 116, No. 3 (2003), p. 820.

64. Brenda Danet, 'Text as mask: gender, play and performance on the Internet', in Steve Jones (ed.) *Cybersociety 2.0: Computer-Mediated Communication and Community Revisited*, Thousand Oaks, CA: Sage, 1998, pp. 129–58; Michele Rodino, 'Breaking out of binaries: reconceptualising gender and its relationship to language in computer-mediated communication', *Journal of Computer-Mediated Communication*, Vol. 3, No. 3 (1997) [online].

65. Peter Kollock and Marc Smith, 'Introduction: communities in cyberspace', in Marc Smith and Peter Kollock (eds), *Communities in Cyberspace*, London: Routledge, 1999, p. 11.

66. Pierre Bourdieu, *Masculine Domination*, Stanford: Stanford University Press, 2001, p. 53.

67. Ibid, p.106. In the academic field (for example) there is a gendered opposition between dominant disciplines, law and medicine, and dominated disciplines, the sciences and the humanities, and within the latter group, between the sciences, with everything that is described as 'hard', and the humanities ('soft'); or again, between sociology, situated on the side of the agora and politics, and psychology, which is condemned to interiority, like literature. Ibid, p. 105.

68. Judy Wajcman, *Techno Feminism*, Polity Press, Cambridge, 2004, p. 77.

69. Ann Gray, *Video Playtime: The Gendering of A Leisure Technology*, London: Routledge, 1992.

70. Elizabeth K. Kelan, 'Tools and toys: communicating gendered positions towards technology', *Information, Communication and Society*, Vol. 10, No. 3 (June 2007), p. 376.
71. Graeme Kirkpatrick, *Critical Technology: A Social Theory of Personal Computing*, Aldershot, UK: Ashgate, 2004.
72. Tiziana Terranova, 'Free labour: producing culture for the digital economy', *Social Text*, Vol. 18, No. 2 (1999), p. 49.
73. Dale Spender, *Nattering on the Net: Women, Power and Cyberspace*, Spinifex, Canada, 1995, p. 196.
74. Susan C. Herring, 'Gender and power in online communication', in Janet Holmes and Miriam Meyerhoff (eds), *The Handbook of Language and Gender*, Oxford: Blackwell, 2003, pp. 202–28.
75. Susan C. Herring, 'Bringing familiar baggage to the new frontier: gender differences in computer-mediated communication', in Victor Vitanza (ed.), *CyberReader*, Boston: Allyn & Bacon, 1996, pp. 144–54.
76. Susan C. Herring, 'Two variants of an electronic message schema', in Susan C. Herring (ed.), *Computer-Mediated Communication: Linguistic, Social and Cross-Cultural Perspectives*, Amsterdam: John Benjamins, 1996, p. 104.
77. Susan C. Herring, 'Gender and democracy in computer-mediated communication', *Electronic Journal of Communication*, Vol. 3, No. 2 (1993) [online].
78. Alexanne Don, 'The dynamics of gender perception and status in email-mediated group interaction V1.0', *Trans/forming Cultures eJournal*, Vol. 2, No. 2 (December 2007), pp. 120–1 [online].
79. Herring, 'Bringing familiar baggage'.
80. Susan C. Herring, Kirk Job-Sluder, Rebecca Scheckler and Sasha Barab, 'Searching for safety online: managing "trolling" in a feminist forum', *The Information Society*, No. 18 (2002), pp 371–84.
81. Ibid.
82. Susan C. Herring, 'The rhetorical dynamics of gender harassment on-line', *The Information Society*, Vol. 15, No. 3 (1999), pp. 151–67.
83. Herring et al., 'Searching for safety online'.
84. Skeggs, 'Exchange, value and affect', p. 85.

Chapter 4

1. Craig Calhoun, 'Habitus, field and capital', in Craig Calhoun, Edward LiPuma and Moishe Postone (eds), *Bourdieu: Critical Perspectives*, Chicago: University of Chicago Press, 1993, p. 82.

2. Luc Boltanski and Laurent Thévenot, *On Justification: Economies of Worth*, Princeton, NJ: Princeton University Press, 2006 [1991].

3. Harold Garfinkel, 'Studies of the routine grounds of everyday activities', in *Studies in Ethnomethodology*, Englewood Cliffs, NJ: Prentice Hall, 1969, p. 37.

4. Luc Boltanski, *L'Amour et la justice comme compétences: Trois essais de sociologie de l'action*, Paris: Métailié, 1990, p. 37.

5. Michel De Certeau, *Heterologies: Discourses on the Other*, Manchester: Manchester University Press, 1986, p. 183.

6. Pierre Bourdieu and Loïc Wacquant, *An Invitation to Reflexive Sociology*, Chicago: University of Chicago Press, 1992, p. 113.

7. Herbert Blumer, *Symbolic Interactionism: Perspective and Method*, Englewood Cliffs, NJ: Prentice Hall, 1969.

8. Boltanski and Thévenot, *On Justification*.

9. Luc Boltanski and Laurent Thévenot, 'The sociology of critical capacity', *European Journal of Social Theory*, Vol. 2, No. 3, 1999, pp. 359–78.

10. Michael Walzer, *Spheres of Justice*, New York: Basic Books.

11. See Jacques Bidet, 'L'esprit du capitalisme: Questions à Luc Boltanski et Eve Chiappello', in Jean Lojkine (ed.), *Les Sociologies critiques du capitalisme*, Presses Universitaires de France, 2002, pp. 215–33.

12. The notion that people use toolkits or repertoires of action has also been put forward by others. Lamont speaks of symbolic repertoires as tools of social inclusion and exclusion. See Michèle Lamont, *Money, Morals, and Manners*, Chicago: University of Chicago Press, 1992. Ann Swidler argues that it is not ideas that matter, but strategies of action such as behavioural habits, styles and skills. Though agreeing that people have reflexivity, she is less optimistic as to the soundness of their judgment: 'In looking at the more-or-less reasoned arguments people make, we see them drawing from their repertoires, trying various rationales with little concern about coherence.' Hence the reference to a mixed bag of justificatory tools used on an ad hoc basis. If there are contradictory causes for action, the true springs for action must lie outside peoples' subjectivities. According to Swidler these springs are found in institutions, which reward conformity or deviance from usual patterns of action, and later require the sense-making mechanism of moral justification. See Ann Swidler, *Talk of Love: How Culture Matters*, Chicago: University of Chicago Press, 2001, p. 34.

13. Peter Wagner, 'After justification: repertoires of evaluation and the sociology of modernity', *European Journal of Social Theory*, Vol. 2, No. 3 (1999), p. 347.

14. Mary Douglas and Steven Ney, *Missing Persons: A Critique of Personhood in the Social Sciences*, Berkeley: University of California Press, 1998.

15. David Beetham, *The Legitimation of Power*, London: Macmillan, 1991.

16. Jean-Jacques Rousseau, *The Social Contract*, London: Penguin Classics, 1968.

17. Chip Salzenberg, 'What is Usenet?', communication posted to news.admin, 1991. Cited in Bryan Pfaffenberger, '"If I want it, it's OK": Usenet and the (outer) limits of free speech', *The Information Society*, Vol. 12, Issue 4 (1996), pp. 365–86.

18. Albert Langer, 'Re: What Is Usenet', communication posted to news.admin, 1991. Quoted in Pfaffenberg, '"If I Want it, it's OK"', p. 378.

19. Pfaffenberg, '"If I Want it, it's OK"', p. 380.

20. The conflict also highlights a common problem in anti-authoritarian groups, the emergence of non-democratic ruling cliques. In an essay on informal power hierarchies in the feminist movement of the early 1970s, Jo Freeman remarked that a laissez-faire group is about as realistic as a laissez-faire society: the idea becomes a smokescreen for the strong or the lucky to establish unquestioned hegemony over others. Members of friendship cliques or 'elites' in collectivist groups will listen more attentively to peers, and interrupt them less; they will repeat each others' points and give in amiably. This explains why it is important for anti-authoritarian groups to have explicit or formal structures: because rules are then known to all. Without such structures, rules are known only to a few, whilst the majority remain in confusion and ignorance of what is happening. Criteria for membership in friendship groups are personal qualities such as a person's background, personality or allocation of time, rather than competence, dedication to the cause, talents or potential contribution. Elites contradict not only sovereign collective authority, but also the basis of charismatic authority. To pre-empt the emergence of elites, Freeman recommends adopting the following measures: (1) delegation of specific authority to specific individuals for specific tasks by democratic procedure; (2) responsibility to the group; (3) distribution as much as possible; (4) rotation of tasks among individuals; (5) allocation of tasks according to ability, or training up of people; (6) diffusion of information; (7) equal access to resources. See Jo Freeman, 'The tyranny of structurelessness', *Berkeley Journal of Sociology*, Vol. 17, 1972–73.

21. Joseph Raz, *The Concept of a Legal System: An Introduction to the Theory of the Legal System*, 2nd edn, Oxford: Clarendon Press, 1980.

22. Jennifer L. Mnookin, 'Virtual(ly) law: the emergence of law in LambdaMOO', *Journal of Computer-Mediated Communication*, Vol. 2, No. 1 (1996) [online].

23. James Boyle, 'Foucault in cyberspace: surveillance, sovereignty and hardwired censors', *University of Cincinnati Law Review*, Vol. 66 (1997), pp. 177–205, p. 188).

24. Lawrence Lessig, *Code and Other Laws of Cyberspace*, New York: Basic Books, 1999, p. 206.

25. Ibid, p. 5. If the protological principles of the Internet, such as peer-to-peer distribution and the equality of all digital packets before the router, were damaged or repressed, then Net neutrality would cease to exist. However, capitalist interests are in fact divided: content producers want strict policing of copyright, while it is in the interest of hardware vendors that people feel they have access to 'free' or reproducible content.

26. Carleton Kemp Allen, *Law in the Making*, Oxford University Press, 1964.

27. David G. Post and David R. Johnson, 'Law and borders: the rise of law in cyberspace', *Stanford Law Review*, Vol. 48 (1995), pp. 1367–402.

28. Benjamin Wittes, 'Witnessing the birth of a legal system', *The Connecticut Law Tribune*, 27 February 1995, p. 8A.

29. Henry Perritt, 'Cyberspace self-government: town-hall democracy or rediscovered royalism?', *Berkeley Technology Law Journal*, Vol. 12 (1997), p. 426.

30. Ibid, p. 456.

31. Ibid.

32. Robert Ellickson, *Order Without Law: How Neighbours Settle Disputes*, Cambridge, MA: Harvard University Press, 1991.

33. Ibid, p. 167.

34. Robert Cooter, 'Review: against legal centrism', *California Law Review*, Vol. 81, No. 1 (1993), p. 420.

35. Elinor Ostrom, *Governing the Commons: The Evolution of Institutions for Collective Action*, New York: Cambridge University Press, 1990, p. 154. Ostrom formulates a series of necessary conditions for successful self-organised communities: there needs to be congruence between group norms and local conditions; collective choice-arrangements; monitoring procedures; conflict-resolution mechanisms; clearly defined community relationships; graduated sanctions; and local enforcement of local rules. Empirical observations of self-organised groupings collectively managing common resources (such as fishing communities) have confirmed the model's validity. This approach offers important insights, and

wherever sovereign authority is strongly present in online tribes we will find a combination of these traits. However, Ostrom's model cannot account for the persistence of online charismatic authority. For example, though Wikipedia demonstrates most of the self-governing traits outlined above, its founder's prestige allows him periodically to override these collective arrangements in an undemocratic (though legitimate) manner.

36. Patrick B. O'Sullivan and Andrew J. Flanagin, 'Reconceptualising "flaming" and other problematic messages', *New Media & Society*, Vol. 5, No. 1 (2003), pp. 69–94.
37. Ibid, p. 84.
38. Patricia G. Lange, 'What is your claim to flame?', *First Monday*, Vol. 11, No. 9 (September 2006) [online].
39. Douglas W. Maynard, 'How children start arguments', *Language in Society*, Vol. 14 (1985), pp. 1–30. [Cited in Lange, 'Claim to flame'].
40. Thomas Benatouïl, 'A tale of two sociologies: the critical and pragmatic stance in contemporary French sociology', *European Journal of Social Theory*, Vol. 2, No. 3 (1999), p. 390.
41. Tamir Maltz, 'Customary law and power in Internet communities', *Journal of Computer-Mediated Communication*, Vol. 2, No. 1 (June 1996).
42. Richard C. MacKinnon, 'Punishing the persona: correctional strategies for the virtual offender', in Steve Jones (ed.), *Virtual Culture: Identity and Communication in Cybersociety*, Thousand Oaks, CA: Sage, 1997, pp. 206–35.
43. John R. Suler and Wende Phillips, 'The bad boys of cyberspace: deviant behaviour in multimedia chat communities', *Cyberpsychology and Behaviour*, No. 1 (1998), pp. 275–94.
44. Perritt, 'Cyberspace self-government', p. 424.
45. Mark A. Lemley, 'The law and economics of Internet norms', *Chicago-Kent Law Review*, Vol. 73 (1997), p. 1271.
46. Sally Hambridge, 'RFC 1855 Netiquette Guidelines', Network Working Group, October 1995 [online].
47. Ibid.
48. Ibid.
49. Lemley, 'The law and economics of Internet norms', p.1262.
50. Susan C. Herring, Kirk Job-Sluder, Rebecca Scheckler and Sasha Barab, 'Searching for safety online: managing "trolling" in a feminist forum', *The Information Society*, No. 18 (2002), p. 378.
51. Ibid, p. 380.
52. Wittes, 'Witnessing the birth of a legal system', p. 8A.

53. The aversion towards censorship explains the outrage felt by the Usenet community towards the Church of Scientology (COS) in the 1990s. The COS attempted to erase postings by a user known as 'cancelpoodle' and also argued that any discussion of Scientology on Usenet constituted an intellectual property offence as the term itself is copyrighted.
54. Lemley, 'The law and economics of Internet norms', p. 1285.
55. Ibid.
56. Perritt, 'Cyberspace self-government', p. 442.
57. Mnookin, 'Virtual(ly) law'.
58. Julian Dibbell, 'A rape in cyberspace: or how an evil clown, a Haitian trickster spirit, two wizards, and a cast of dozens turned a database into a society', *The Village Voice*, 21 December 1993, pp. 36–42.
59. Pavel Curtis, 'LambdaMOO takes a new direction' [online].
60. Mnookin, 'Virtual(ly) law'.
61. Ibid.
62. Ibid.
63. Ibid.

Chapter 5

1. 'What is primitivism?', *Primitivism.com* [online].
2. Theodore Kaczynski, 'Thesis 94', *Industrial Society and its Future*, 1995 [online].
3. Pierre Bourdieu, *Distinction: A Social Critique of the Judgement of Taste*, Cambridge, MA: Harvard University Press, 1984 [1979].
4. Since 1996 the Internet Archive has been saving copies of websites. These archives can be consulted through the Wayback Machine.
5. Christian Fuchs, *Internet and Society: Social Theory in the Information Age*, London and New York, Routledge, 2008.
6. The only exception being a (broken) link to Kaczynski's *Industrial Society*; Kaczynski's name was also removed from the list of authors, perhaps indicating a desire on the part of the site's owner to avoid the 'activist' label.
7. John Zerzan, 'Why Primitivism?', *Telos*, No. 124 (2002), p. 171.
8. Ibid, p. 167.
9. Fredy Perlman, *Against His-Story, Against Leviathan*, Detroit: Black and Red, 1983; David Watson, *Against the Megamachine: Essays on Empire and Its Enemies*, Brooklyn, Autonomedia/Fifth Estate, 1997.

10. Langdon Winner, *The Whale and the Reactor: A Search for Limits in an Age of High Technology*, Chicago: University of Chicago Press, 1986; Jacques Ellul, *The Technological Bluff*, Grand Rapids, MI: William B. Eerdmans, 1990; Kirkpatrick Sale, *Rebels Against the Future: The Luddites and Their War on the Industrial Revolution – Lessons for the Computer Age*, New York: Addison-Wesley, 1995.
11. Edward Abbey, *The Monkey Wrench Gang*, Philadelphia: Lippincott, 1975.
12. Arne Naess, *Ecology, Community and Lifestyle: Outline of an Ecosophy*, Cambridge: Cambridge University Press, 1990.
13. William Devall and George Sessions, *Deep Ecology*, Salt Lake City: Gibbs M. Smith, 1985.
14. Bron Taylor, 'Earth First! and global narratives of popular ecological resistance', in *Ecological Resistance Movements: The Global Emergence of Radical and Popular Environmentalism*, Bron Taylor (ed.), Albany: State University of New York Press, 1995, p. 16.
15. Ariel Salleh and Meira Hansen, 'On production and reproduction, identity and non-identity in ecofeminist theory', *Organization & Environment*, No. 12 (1999), pp. 207–18.
16. Russell Goldman, 'Environmentalists classified as terrorists get stiff sentences', *ABC News*, 25 May 2007 [online].
17. Marshall D. Sahlins, *Tribesmen*, Englewood Cliffs, NJ: Prentice-Hall, 1968, p. 8.
18. Murray Bookchin, *Post-Scarcity Anarchism*, London: Wildwood House, 1971, p. 78.
19. Murray Bookchin, 'Social ecology vs. deep ecology: a challenge for the ecology movement', *Green Perspectives*, No. 4–5, 1987, pp. 3–18.
20. Ibid., p. 16.
21. Murray Bookchin, *Social Anarchism or Lifestyle Anarchism: An Unbridgeable Chasm*, San Francisco: AK Press, 1995.
22. Ibid., p. 27.
23. Ibid., p. 29.
24. David Watson, *Beyond Bookchin: Preface for a Future Social Ecology*, Brooklyn: Autonomedia, 1996.
25. Murray Bookchin, 'Whither anarchism? a reply to recent anarchist critics', in *Anarchism, Marxism and the Future of the Left*, San Francisco: AK Press, 1999, pp. 160–86.
26. Ibid., p. 172.
27. Bob Black, 'Withered anarchism: a surrebuttal to Murray Bookchin', *Primitivism.com*, 1999 [online].

28. Edwin Wilmsen, *Land Filled with Flies: A Political Economy of the Kalahari*, Chicago: University of Chicago Press, 1989.
29. Bob Black, 'Book filled with lies', *Primitivism.com*, 1999 [online].
30. Ken Knabb, *Public Secrets*, Berkeley: Bureau of Public Secrets, 1997.
31. Ken Knabb, 'The poverty of primitivism', *Bureau of Public Secrets*, 2001 [online].
32. John Filiss, 'Intellectual impoverishment', *Primitivism.com*, 2001 [online].
33. In order to demonstrate anti-establishment credentials, the discourse used on such forums is highly colloquial. One randomly selected thread included the following: 'you sound like a real dope', 'you fucking dipshit', 'pencil-dick', etc. Various authors, 'Announcing the eighth annual BASTARD conference', *Antipolitics forum*, 23 January 2008 [online].
34. Bookchin, 'Whither anarchism?', p. 169.
35. Ken Knabb, 'A look at some of the reactions to "Public Secrets"', *Bureau of Public Secrets*, 2000 [online].
36. John Zerzan, '*Willful Disobedience*: more radical than thou?', *Green Anarchy*, No. 7 (2001), p. 21.
37. Knabb, 'A look at some of the reactions'.
38. Knabb, *Public Secrets*, p. 84.
39. Black, 'Withered anarchism'.
40. Ibid.
41. Ibid.
42. Kaczynski, *Industrial Society*, Thesis 161.
43. Val Burris, Emery Smith and Anne Strahm, 'White supremacist networks on the Internet', *Sociological Focus*, Vol. 33, No. 2 (2000), p. 232.
44. Manuel Castells, *The Power of Identity* (Vol. 2 of *The Information Age: Economy, Society and Culture*), 2nd edn, London: Blackwell, 2004, p. 94.
45. Scott Lash, *Critique of Information*, London: Sage, 2002.
46. John Connor et al., 'An Open Letter on Technology and Mediation', *Primitivism.com*, 1999 [online].
47. Ibid.

Chapter 6

1. Mathieu O'Neil, 'The lay of the land: portrait of the zinester as a social statistic', *Factsheet 5*, No. 62 (1997), pp. 10–11; Mathieu O'Neil, 'Les paradoxes d'une dissidence Californienne: les "zines"

de la San Francisco Bay Area (1980–1995)', in Annick Foucrier and Antoine Coppolani (eds), *La Californie: périphérie ou laboratoire?* Paris: L'Harmattan, 2004, pp. 47–59.

2. Stephen Duncombe, *Notes from Underground: Zines and the Politics of Alternative Culture*, London and NewYork: Verso, 1997.

3. Meteor Blades, 'Open Thread for Night Owls & Early Birds', *Daily Kos*, 8 May 2008 [online].

4. Matt Stoller, 'What is this new movement', *TPMCafe*, 15 January 2007 [online].

5. Markos Moulitsas, 'Day 1', *Daily Kos*, 26 May 2002 [online].

6. Yochai Benkler, *The Wealth of Networks: How Social Production Transforms Markets and Freedom*, Yale: Yale University Press, 2006.

7. See Ben Bagdikian, *The New Media Monopoly*, Boston: Beacon Press, 2004; Robert W. McChesney, *Rich Media, Poor Democracy: Communication Politics in Dubious Times*, New York: New Press, 2000.

8. Tanni Haas, 'From public journalism to the public's journalism? Rhetoric and reality in the discourse on weblogs', *Journalism Studies*, Vol. 6, No. 3 (2005), p. 389.

9. Lakshmi Chaudhry, 'Can blogs revolutionise progressive politics?', *In These Times*, February 2006 [online].

10. Stephen Coleman, 'Blogs and the new politics of listening', *Political Quarterly*, Vol. 76, No. 2 (2005), p. 274.

11. Rebecca Blood, 'Weblogs: a history and perspective', *Rebecca's Pocket*, 7 September 2000 [online].

12. Manuel Castells, *The Power of Identity* (Vol. 2 of *The Information Age: Economy, Society and Culture*), 2nd edn, London: Blackwell, 2004, p. 370.

13. Eszter Hargittai, Jason Gallo and Matthew Kane, 'Cross-ideological discussions among conservative and liberal bloggers', *Public Choice*, Vol. 134, Nos. 1–2 , pp. 67–86.

14. Markos Moulitsas, 'The rise of open source', *Daily Kos*, 11 February 2005 [online].

15. See Gary Wolf, 'How the Internet invented Howard Dean', *Wired*, Vol. 12, No. 1 (January 2004) [online].

16. Quoted in Terry McDermott, 'Blogs can top the presses', *LA Times*, 17 March 2007, p. A1.

17. The archetypal example of distributed fact-checking punctured CBS's investigative news show *60 Minutes* in 2004. Documents which, in CBS's view, showed irregularities in George W. Bush's National Guard service were found to be fakes. Starting from the

conservative weblog *Free Republic*, this information travelled to the conservative information and gossip clearing house *The Drudge Report* and from then on into the traditional media. A day after they were broadcast on CBS, ABC disputed the credibility of the accusations.

18. Henry Farrell, 'Bloggers and parties: can the Netroots reshape American democracy?', *Boston Review*, September/October 2006 [online].

19. Ibid.

20. Henry Farrell and Daniel W. Drezner, 'The power and politics of blogs', *Public Choice*, Vol. 134, Nos. 1–2 (2008), p. 22.

21. Michael C. Munger, 'Blogging and political information: truth or truthiness?', *Public Choice*, Vol. 134, Nos. 1–2 (2008), pp. 125–38.

22. Benjamin Wallace-Wells, 'Kos call', *Washington Monthly*, January/February 2006 [online].

23. Daily Kos relies on a news management software named Scoop which allows users to vet others' stories and comments. It was originally invented by Rusty Foster, creator of the (Slashdot-inspired) kuro5hin website.

24. On 4 June 2008, the day after Barack Obama secured the Democratic nomination, 595 diaries were posted to Daily Kos – approximately every 2.5 minutes on average.

25. Bruce Bimber, *Information and American Democracy: Technology in the Evolution of Political Power*, Cambridge: Cambridge University Press, 2003.

26. Kagro X, 'Some thoughts on journalism and the blogosphere', *Daily Kos*, 12 May 2007 [online].

27. BlueBeliever, 'Who are Georgia10, Armando, Darksyde, SusanG, Mcjoan, etc.?', *Daily Kos*, 28 March 2006 [online].

28. Various authors, comments to BlueBeliever, 'Who are Georgia10, Armando, Darksyde, SusanG, Mcjoan, etc.?' [online].

29. Ibid.

30. Ibid.

31. Cass Sunstein, *Republic.com*, Princeton: Princeton University Press, 2001.

32. Hunter, 'The tao of troll rating', *Daily Kos*, 26 May 2006 [online].

33. Ibid.

34. 'Who speaks for this site?', *Daily Kos FAQ* [online].

35. Sunstein, *Republic.com*.

36. See Lada Adamic and Natalie Glance, 'The political blogosphere and the 2004 U.S. election: divided they blog', in *LinkKDD*

'05: *Proceedings of The Third International Workshop on Link Discovery*, Chicago, IL, 21–25 August 2005, pp. 36–43; Mark Tremayne, Nan Zheng, Jae Kok Lee and Jaekwan Jeong, 'Issue publics on the Web: applying network theory to the war blogosphere', *Journal of Computer-Mediated Communication*, Vol. 12, No. 1 (2006) [online].

37. George Packer, 'The revolution will not be blogged', *Mother Jones*, May/June 2004 [online].

38. Hargittai et al., 'Cross-ideological discussions', p. 85.

39. Bill O'Reilly, 'The Factor', *Fox News*, 13 May 2008.

40. Budhydharma, 'YOU can't handle the truth! Trolls and troll hunting', *Daily Kos*, 11 May 2006 [online].

41. Gérard Chaliand, *Anthologie mondiale de la stratégie*, Paris: Robert Laffont, 1990, p. xvi.

42. Susan C. Herring and John Paolillo, 'Gender and genre variation in weblogs', *Journal of Sociolinguistics*, Vol. 10, No. 4 (2006), pp. 439–59; Dustin Harp and Mark Tremayne, 'The gendered blogosphere: examining inequality using network and feminist theory', *Journalism and Mass Communication Quarterly*, No. 83 (Summer 2006), pp. 247–64; Clancy Ratliff, 'Attracting readers: sex and audience in the blogosphere', *Scholar and Feminist Online*, Vol. 5, No. 2 (Spring 2007) [online].

43. Jose Antonio Vargas, 'A diversity of opinions, if not opinionators', *Washington Post*, 6 August 2007, p. C01.

44. 'Pie Fight', *Daily Kos* [online].

45. Progressive South, 'Why the blogosphere white-out of Jena 6?', *Daily Kos*, 20 September 2007 [online]; dnA, 'Radio Silence', *Too Sense*, 25 April 2008 [online].

46. Pam Spaulding, 'Diversity in the blogosphere 2.0', *Daily Kos*, 21 September 2006 [online].

47. Markos Moulitsas, 'Clinton trashes "activists" and MoveOn at closed-door fundraiser', *Daily Kos*, 18 April 2008 [online].

48. And what of *my* cohorts, the 'Generation Xers'? Those cynical underachievers were forgotten, it seems.

49. Alegre, 'Does this guy speak for Obama?', *Daily Kos*, 14 December 2007 [online].

50. Markos Moulitsas, 'Taking a hard line on ratings war', *Daily Kos*, 27 December 2007 [online].

51. Various authors, comments to Delaware Dem, 'The Hillary Roll Call (Updated)', *Daily Kos*, 28 December 2007 [online].

52. Ibid.

53. See Robin Givhan, 'Hillary Clinton's tentative dip into new neckline territory', *Washington Post*, 20 July 2007, p. C01; Joan Vennochi,

'That Clinton cackle', *Boston Globe*, 30 September 2007, p. A1; Robin Marie Cocco, 'Misogyny I won't miss', *Washington Post*, 15 May 2008, p. A15; Robin Morgan, 'Goodbye to all that', *Guardian Online*, 14 February 2008 [online].

54. Michelle Goldberg, '3 A.M., for Feminism: Clinton dead-enders and the crisis in the women's movement', *New Republic*, 6 June 2008 [online].

55. Rena RF, 'Don't make me stop this car and come back there!', *Daily Kos*, 25 February 2008 [online].

56. Slinkerwink, 'Mothership: Clinton tax returns out $109 million!', *Daily Kos*, 4 April 2008 [online].

57. Todd Beeton, 'Voices of reason amongst the derangement', *MyDD*, 4 April 2008 [online].

58. Alegre, 'Hillary & International Women's Day', *Daily Kos*, 8 March 2008 [online].

59. Alegre, 'Writers strike at Daily Kos', *Daily Kos*, 14 March 2008 [online].

60. Markos Moulitsas, 'The Clinton civil war', *Daily Kos*, 17 March 2008 [online].

61. Goldberry, 'Obamaphiles carry out jihad on Daily Kos', *Daily Kos*, 14 January 2008 [online].

62. Riverdaughter, 'An invitation to Kossacks in exile', *The Confluence*, 26 January 2008 [online].
This post was removed in early May (it was last cached by Google on May 1) and was replaced by a post entitled 'An invitation to Democrats in exile' [online]. The original post is still visible via the Internet Archive's Wayback Machine.

63. Gabriele Droz, comment to Riverdaughter, 'An invitation to Kossacks in exile' [online].

64. James Wolcott, 'When Democrats go post-al', *Vanity Fair*, June 2008 [online].

65. Jeralyn, 'Wolcott on progressives "going postal"', *TalkLeft*, 30 April 2008 [online].

66. Various authors, comments to Riverdaughter, 'An invitation to Kossacks in exile' [online].

67. Various authors, comments to Jeralyn, 'Wolcott on progressives "going postal"' [online].

68. Various authors, comments to Big Tent Democrat, 'About those proud Hillary haters', *TalkLeft*, 18 April 2008 [online].

69. Jerome Armstrong, 'The Vote', *MyDD*, 12 February 2008 [online].
Needless to say, this did not go down very well with Obama supporters. On 22 March, a Daily Kos diary alleged that Armstrong was committing the sin of sins: being dishonest with the electoral

arithmetic by writing that the primary states of North Carolina and Oregon, to be decided in May, and generally thought to be solidly pro-Obama, were 'toss-ups'. The commenters were outraged: 'It is just sad because it has nothing to do with reality and he used to be a respected analyst'; 'The place feeds on intellectual dishonesty'; 'MyDD has become a shrill, Kool Aid hack site of cultists.' Various authors, comments to NMLib, 'Jerome Armstrong's intellectual dishonesty', *Daily Kos*, 22 March 2008 [online].

70. Hillarywillwin, comment to Alegre, 'Is This What We've Been Reduced to?', *MyDD*, 10 May 2005 [online].

71. Alegre, 'OK, so tell us what IS allowed', *MyDD*, 19 May 2008 [online].

72. Chrisblask, 'What IS allowed: discussion', comment to Alegre, 'OK, so tell us what IS allowed' [online].

73. Various authors, comments to Alegre, 'Trouble in the heartland', *MyDD*, 27 May 2008 [online].

Chapter 7

1. A, 'Just a simple question for the candidate', *debian-vote*, 3 March 2004 [online].

2. MS, 'Re: Just a simple question for the candidate', *debian-vote*, 4 March 2004. Sexism in the Linux community and strategies to address sexist behaviour had previously been addressed in Val Henson, *HOWTO Encourage Women in Linux*, 29 October 2002 [online].

3. Pierre Bourdieu, *Masculine Domination*, Stanford: Stanford University Press, 2001.

4. Eric Raymond, *The Cathedral and the Bazaar: Musings on Open Source and Linux by an Accidental Revolutionary*. Sebastopol, CA: O'Reilly, 1999.

5. Ian Murdock, 'Debian: a brief retrospective', *LinuxPlanet*, 15 August 2003 [online].

6. Frauke Lehmann, 'FLOSS developers as a social formation', *First Monday*, Vol. 9, No. 11 (November 2004) [online].

7. Martin Krafft, interviewed by Sal Cangeloso, *geek.gom*, 16 January 2006 [online].

8. Eben Moglen, 'Anarchism triumphant: free software and the death of copyright', *First Monday*, Vol. 4, No. 8 (August 1999) [online].

9. Cited in Moglen, 'Anarchism triumphant'.

10. Manoj Srivastava, 'Why Debian?', *Debian Wiki*.

11. Ibid.

12. Martin Krafft, *The Debian System*, San Francisco, CA: No Starch Press, 2005, p. 18.

13. 'General Resolution: Why the GNU Free Documentation License is not suitable for Debian main', *Debian* [online].

14. '2.1. General rules', *Debian Constitution* [online].

15. Jean-Antoine-Nicolas de Caritat, Marquis de Condorcet, *Essai sur l'application de l'analyse à la probabilité des décisions rendues à la pluralité des voix*, New York: Chelsea Publishing Company, 1972 [1785].

16. Nicolas Auray, 'Le Modèle souverainiste des communautés en ligne: l'impératif participatif et la désacralisation du vote', *Hermès*, No. 47 (2007).

17. Krafft, *The Debian System*, p. 50.

18. Nicolas Auray, 'Le Sens du juste dans un noyau d'experts: Debian et le puritanisme civique', in Bernard Conein, Francoise Massit-Folléa and Serge Proulx (eds), *Internet, une utopie limitée: Nouvelles régulations, nouvelles solidarités*, Laval, Quebec: Presses de l'Université Laval, 2005.

19. Nicolas Auray, 'The decision-making process within a huge project: discussions and deliberations in Debian', Libre/Open Source Software: Which Business Model? Calibre Workshop, Paris, 4 March 2005.

20. Krafft, *The Debian System*, p. 53.

21. Josh Lerner and Jean Tirole, 'Some simple economics of open source', *Journal of Industrial Economics*, Vol. 50, No. 2 (2002), pp. 194–327.

22. Audris Mockus, Roy Fielding and James Herbsleb, 'A case study of open source software development: the Apache server', in *Proceedings of the Twenty-second International Conference on Software Engineering*, Limerick, Ireland, ACM Press, 2000, pp. 263–72.

23. Krafft, *The Debian System*, p. 54.

24. Bernard Conein, 'Communauté épistémique et réseaux cognitifs: coopération et cognition distribuée', *Revue d'Economie Politique*, No. 113 (2004), pp. 141–59.

25. Bernard Conein, 'Relations de conseil et expertise collective: comment les experts choisissent-ils leurs destinataires dans les listes de discussion?', *Recherches Sociologiques* Vol. 35, No. 3 (2004), pp. 61–74.

26. Auray, 'Le Sens du juste'.

27. Siobhán O'Mahony and Fabrizio Ferraro, 'The emergence of governance in an open source community', *Academy of Management Journal*, Vol. 50, No. 5 (2007), pp. 1079–106.

28. Siobhán O'Mahony and Fabrizio Ferraro, 'Managing the boundary of an "open" project', Workshop on the Network Construction of Markets, Santa Fe Institute, 2003.

29. Ibid.

30. Auray, 'Le Modèle souverainiste'.

31. '5.3. Project Leader: Procedure', Debian Constitution [online].

32. Ibid.

33. SH, 'Re: Making Debian work: a question of trust indeed', debian-project, 21 November 2007 [online].

34. 'General Resolution: endorse the concept of Debian Maintainers', Debian [online].

35. Lehmann, 'FLOSS developers as a social formation'.

36. Richard Stallman, 'Copyright and globalisation in the age of computer networks', in Rishab Aiyer Ghosh (ed.), CODE: Collaborative Ownership and the Digital Economy, Cambridge, MA: MIT Press, 2005, pp. 317–35.

37. Ian Murdock, 'Ubuntu vs. Debian, reprise', Ian Murdock's Weblog, 20 April 2005 [online].

38. 'Community: Debian and Ubuntu', Ubuntu [online].

39. Matthew Garrett, 'I resigned from Debian today', LiveJournal, 28 August 2006 [online].

40. MP, 'Re: "I do consider Ubuntu to be Debian", Ian Murdock', debian-user, 19 March 2007 [online].

41. Scott James Remnant, 'Having left Debian', Netsplit, 2 September 2006 [online].

42. Martin Krafft, 'Ubuntu and Debian', Madduck, 2006 [online].

43. Joey Hess, 'The supermarket thing', 21 October 2007 [online].

44. Sam Varghese, 'A conversation with Martin Michlmayr', IT Wire, 31 January 2008 [online].

45. Auray, 'Le Sens du juste'.

46. Robert K. Merton, The Sociology of Science: Theoretical and Empirical Investigations, Chicago: University of Chicago Press, 1973.

47. Various authors, 'Re: Critique constructive de Debian', debian-devel-french, 29 September 2007 [online].

48. SL, 'Re: [Yaird-devel] Re: Bug#345067: ide-generic on powerpc', debian-ctte, 9 March 2006 [online].

49. Ibid.

50. IJ, 'Flamewars and uncooperative disputants, and how to deal with them', debian-ctte, 9 March 2006 [online].

51. AS, 'Question for all candidates about developer behaviour and abuse', debian-vote, 7 March 2006. [name changed] [online]

52. FP, 'Re: The powerpc port should be removed from etch release candidates', *debian-release*, 1 May 2006 [online].
53. Various authors, 'Re: Issues concerning powerpc and SL', *debian-project*, May 2006 [online].
54. Various authors, 'RE: Public request that action be taken at whoever abused their technical power to remove me from the kernel team at alioth', *debian-project*, May 2007 [online]; various authors, 'A question to the Debian community', *debian-vote*, May 2007 [online].
55. RA, 'Re: A question to the Debian community', *debian-vote*, 10 May 2007 [online].
56. IJ, 'Social committee proposal', *debian-project*, 1 June 2007 [online].
57. AT, 'Re: Technical Committee resolution', *debian-vote*, March 2008 [online].
58. Ibid.
59. Ibid.
60. MS, 'Re: Technical Committee resolution', *debian-vote*, 31 March 2008 [online].
61. MS, 'Re: Technical Committee resolution', *debian-vote*, 2 April 2008 [online].
62. RA, 'Re: Deficiencies in Debian', *debian-project*, 10 May 2007 [online].
63. Raphaël Hertzog, 'DSA needs a leader', *Buxy rêve tout haut*, 26 September 2007 [online].

Chapter 8

1. Unless otherwise specified, the English language version is referred to.
2. Cited in Susan L. Bryant, Andrea Forte and Amy Bruckman, 'Becoming Wikipedian: transformations of participation in a collaborative online encyclopaedia', in *Proceedings of the GROUP International Conference on Supporting Group Work*, Sanibel Island, FL, 2005, pp. 1–10.
3. Marshall Poe, 'The Hive', *Atlantic Monthly*, September 2006 [online].
4. Larry Sanger, 'Britannica or Nupedia? The future of free encyclopaedias', *Kuro5hin*, 25 July 2001 [online].
5. Stacy Schiff, 'Know it all', *New Yorker*, 31 April 2006 [online].
6. David Mehegan, 'Bias, sabotage haunt Wikipedia's free world', *Boston Globe*, 12 February 2006 [online].

7. Jimmy Wales, 'The wisdom of crowds', *Observer*, 22 June 2008 [online].
8. Schiff, 'Know it all'.
9. User:Jimbo_Wales, 'Statement of principles', *Wikipedia* [online].
10. Poe, 'The Hive'.
11. 'Five pillars', *Wikipedia* [online].
12. Yochai Benkler and Helen Nissenbaum, 'Commons-based peer production and virtue', *Journal of Political Philosophy*, Vol. 14, No. 4 (2006), p. 398.
13. Andrew Lih, 'Wikipedia as participatory journalism: reliable sources? Metrics for evaluating collaborative media as a news resource', Fifth International Symposium on Online Journalism, Austin, TX, 16–17 April 2004.
14. Andrea Forte and Amy Bruckman, 'Why do people write for Wikipedia? Incentives to contribute in open-content publishing', GROUP 05 workshop, Sustaining Community: The Role and Design of Incentive Mechanisms in Online Systems, Sanibel Island, FL, 2005. A watchlist is a user page which automatically records any modifications to selected Wikipedia articles or pages.
15. Nicholson Baker, 'The charms of Wikipedia', *New York Review of Books*, Vol. 55, No. 4, 20 March 2008 [online].
16. Cited in Schiff, 'Know it all'.
17. David Shay and Trevor Pinch, 'Six degrees of reputation: the use and abuse of online review and recommendation systems', *ScTS Working Paper*, Cornell University, 2005, p. 8.
18. Jaron Lanier, 'Digital Maoism: the hazards of the new online collectivism', *The Edge*, May 2006 [online].
19. 'The Seigenthaler incident', *Wikipedia* [online].
20. Colin Jackson, cited in Andrew Orlowski, 'Take out a subscription to The Register. Then cancel it, and sign it Disgusted Wikipedian', *The Register*, 23 December 2005 [online].
21. John Borland, 'See who's editing Wikipedia: Diebold, the CIA, a campaign', *Wired*, 14 August 2007 [online]. When learning that a Microsoft employee had altered free-software articles, TV comedian Colbert (of 'truthiness' fame) urged his viewers to modify the Wikipedia entry for 'reality' to 'reality has become a commodity'. As a result, *reality had to be locked down*; the 'Colbert effect' had struck again. On his show, Colbert quoted Jimmy Wales: Wales was 'very disappointed to hear that Microsoft was taking that approach', to which the comedian responded: 'Boo hoo, comrade! Open source software is like free trade – and the invisible hand of the market has the mouse now'. Cited in Andrew

Orlowski, 'Wikipedia defends reality', *The Register*, 2 February 2007 [online].

22. Jim Giles, 'Internet encyclopedias go head to head', *Nature*, No. 438, 15 December 2005, pp. 900–1.

23. Encyclopaedia Britannica, 'Fatally flawed: Refuting the recent study on encyclopedic accuracy by the journal *Nature*', March 2006 [online].

24. 'What is Veropedia?', *Veropedia* [online]

25. Fernanda Viégas, Martin Wattenberg and Kushal Dave, 'Studying cooperation and conflict between authors with history flow visualisations', CHI 2004, Vienna, 24–9 April 2004. Automatically displaying the index authority of principal contributors or articles could improve the distributed expertise system. Clay Shirky has proposed imitating the 'activity level' indicator on free-software repository SourceForge to indicate the number of edits to pages. See Clay Shirky and commenters, 'Wikipedia: the nature of authority, and a lazyweb request', *Corante*, 2 January 2005 [online].

26. Each computer connected to the Internet is allocated a unique Internet Protocol (IP) address.

27. See Aniket Kittur, Ed Chi, Bryan A. Pendleton, Bongwon Suh and Todd Mytkowicz, 'Power of the few vs. wisdom of the crowd: Wikipedia and the rise of the bourgeoisie', Twenty-fifth Annual ACM Conference on Human Factors in Computing Systems (CHI 2007), San Jose, CA, 28 April–3 May 2007; Dennis M. Wilkinson and Bernardo A. Huberman, 'Cooperation and quality in Wikipedia', in *Proceedings of the 2007 International Symposium on Wikis*, Montreal, pp. 157–64; Felipe Ortega and Jesus M. Gonzalez Barahona, 'Quantitative analysis of the Wikipedia community of users', in *Proceedings of the 2007 International Symposium on Wikis*, Montreal, pp. 75–86.

28. Brian Butler, Elisabeth Joyce and Jacqueline Pike, 'Don't look now, but we've created a bureaucracy: the nature and roles of policies and rules in Wikipedia', in *Proceedings of the Conference on Human Factors in Computing Systems*, Florence, Italy, 5–10 April 2008, pp. 1101–10.

29. Fernanda B. Viegas, Martin Wattenberg and Matthew Mckeon, 'The hidden order of Wikipedia', Online Communities and Social Computing HCII, 2007, pp. 445–54.

30. Noam Cohen, 'A contributor to Wikipedia has his fictional side', *New York Times*, 5 March 2007, p. C15.

31. User:Essjay, 'Talk page', *Wikipedia*. (The page has been archived on www.wikipedia-watch) [online].

32. Schiff, 'Know it all'.

33. Jimmy Wales, 'Sysop status', *Wikien-I*, 11 February 2003 [online].
34. Seth Anthony, 'Contribution patterns among active Wikipedians: finding and keeping content creators', *Wikimania*, 5 August 2006.
35. Utilisateur: Valérie75, 'Pourquoi je suis partie, pourquoi je suis revenue', *Wikipédia: Bulletin des administrateurs*, 22 March 2007 [online].
36. 'Guide to request for adminship', *Wikipedia* [online].
37. Moira Burke and Robert Kraut, 'Taking up the mop: identifying future Wikipedia administrators', in *Proceedings of the Conference on Human Factors in Computing Systems*, Florence, Italy, 5–10 April 2008, pp. 3441–6.
38. Jimmy Wales. 'Mediation, arbitration', *Wikien-I*, 16 January 2004 [online].
39. 'Arbitration Committee', *Wikipedia* [online].
40. Poe, 'The Hive'.
41. User:Jimbo_Wales, 'My_desysop_of_Zscout', *Wikipedia:Administrator's Noticeboard*, 29 October 2007 [online].
42. 'Bedford and misogyny', *Wikipedia:Administrator's Noticeboard*, 25 July 2008 [online].
43. Aniket Kittur, Bongwon Suh, Bryan A. Pendleton and Ed. H. Chi, 'He says, she says: conflict and coordination in Wikipedia', in *Proceedings of the Conference on Human Factors in Computing Systems*, San José, CA, 28 April–3 May 2007, p. 453.
44. Ibid, p. 455.
45. Benkler and Nissenbaum, 'Commons-based peer production', p. 397.
46. Bryant et al., 'Becoming Wikipedian'.
47. Butler et al., 'Don't look now'.
48. Andrea Forte and Amy Bruckman, 'Scaling consensus: increasing decentralisation in Wikipedia governance', in *Proceedings of the Hawaiian International Conference of Systems Sciences*, January 2008, p. 157.
49. Here is an edited list of guidelines for dealing with newbies: (1) Avoid intensifiers (terrible, dumb, stupid, bad, good) in commentary. (2) Modulate your approach and wording. (3) Avoid sarcasm in edit summaries and on talk pages. (4) Strive to respond in a measured manner. (5) Accept graciously another person's actions or inactions. (6) Acknowledge differing principles and a willingness to reach consensus. (7) Open yourself towards taking responsibility for the resolution of conflicts. (8) Reciprocate where necessary. (9) Listen

actively. (10) Avoid Wikipedia jargon. (11) Avoid using blocks as a first resort. Wikipedia, 'How to avoid being a biter' [online].

50. User:Kransky, conversation with the author.
51. 'Wheel war', *Wikipedia* [online].
52. Wikipedia, 'Administrator intervention against vandalism' [online].
53. Reid Priedhorsky, Jilin Chen, Shyong K. Lam, Katherine Panciera, Loren Terveen and John Riedl, 'Creating, destroying, and restoring value in Wikipedia', in *Proceedings of the International ACM Conference on Supporting Group Work*, Sanibel Island, FL, 4–7 November 2007, pp. 259–68.
54. 'Misuse of process: what is a troll?', *Wikipedia* [online].
55. Paradoxically, a number of Wikipedia admins take part in the discussions on Wikipedia Review.
56. Viégas et al., 'Studying cooperation and conflict'.
57. Shaka UVM, 'Wikipedia is pretty messed up', comment to 'The Register exposes more Wikipedia abuse', *Slashdot*, 7 December 2007 [online].
58. Parker Peters, 'Lesson #2: Procedure vs Content, or "You didn't genuflect deeply enough"', *LiveJournal*, 18 January 2007 [online].
59. User:Durova, 'Arbitration Committee elections/candidate statements', *Wikipedia*, 9 November 2007 [online].
60. User:Durova, 'Indefinite block of User:!!', *Wikipedia:Administrators' Noticeboard/Incidents*, 18 November 2007 [online].
61. User:Durova, 'Unblock with apologies', *Wikipedia:Administrators' Noticeboard/Incidents*, 18 November 2007 [online].
62. User:Blnguyen, 'Oversighted edits', *Wikipedia:Administrators' Noticeboard/Incidents*, 23 November 2007 [online].
63. User:Durova, 'Requests for arbitration/Durova/Evidence', *Wikipedia*, 26 November 2007 [online].
64. Jayjg, 'Missed opportunities to have avoided the Durova case', *Wikien-I*, 30 November 2007 [online].
65. Risker, 'Missed opportunities to have avoided the Durova case', *Wikien-I*, 27 November 2007 [online].
66. User:Durova, 'Requests for arbitration/Durova/Evidence'.
67. Arbitration Committee, 'Request for arbitration/Durova', *Wikipedia*, 1 December 2007 [online].
68. The interplay between sock-zapping admins and admin-tricking socks irresistibly conjures up the notion that all these people are players in a massive multi-player online role-playing game. This *World of Wordcraft* operates on a global stage, on which feats of combat are publicly archived for the world to see: in terms of

audience and fame, this is obviously much better than the gated battlefields of conventional gaming environments. Nicholson Baker makes a similar point: 'Without the kooks and the insulters and the spray-can taggers, Wikipedia would just be the most useful encyclopedia ever made. Instead it's a fast-paced game of paintball'. See Baker, 'The charms of Wikipedia'.

69. User:Jimbo_Wales, 'Comment', *Wikipedia:Administrators' Noticeboard*, 22 November 2007 [online].
70. Larry Sanger, 'One alternative', *Wikien-I*, 5 December 2007 [online].
71. Ibid.
72. Ibid.
73. User:Amerique, 'User Talk: Community recall', *Wikipedia*.
74. User:Jimbo_Wales, 'User Talk: Community recall', *Wikipedia*.
75. Forte and Bruckman, 'Scaling consensus'.
76. Ibid.

Chapter 9

1. Nicolas Auray, 'Le Modèle souverainiste des communautés en ligne: l'impératif participatif et la désacralisation du vote', *Hermès*, No. 47 (2007).
2. See Siobhán O'Mahony and Fabrizio Ferraro, 'The emergence of governance in an open source community', *Academy of Management Journal*, Vol. 50, No. 5 (2007), pp. 1079–106.
3. Paul M. Harrison, 'Weber's categories of authority and voluntary associations', *American Sociological Review*, Vol. 25, No. 2 (April 1960), p. 236.
4. Joyce Rothschild-Whitt, 'The collectivist organisation: an alternative to rational–bureaucratic models', *American Sociological Review*, Vol. 44, No. 4 (August 1979), p. 509.
5. Ibid, pp. 511–12.
6. Paul Hoffman and Susan Harris, 'The Tao of the IETF: a novice's guide to the Internet Engineering Task Force', RFC 4677, FYI 17, September 2006 [online].
7. Robert Braden, Joyce K. Reynolds, Steve Crocker, Vint Cerf, Jake Feinler and Celeste Anderson (1999) '30 years of RFCs', RFC 2555, April 1999.
8. 'What Wikipedia is not', *Wikipedia* [online].
9. O'Mahony and Ferraro, 'The emergence of governance in an open source community'.
10. David Courpasson and Mike Reed, 'Introduction: bureaucracy in the age of enterprise', *Organization*, Vol. 11, No. 1 (2004), p. 6.

11. Jannis Kallinikos, 'The social foundations of the bureaucratic order', *Organization*, Vol. 11, No. 1 (2004), p. 16.
12. Cornelius Castoriadis, *The Bureaucratic Society*, vols. 1 and 2, Athens: Ypsilon, 1985; Fridrich Hayek, *The Constitution of Liberty*, London: Routledge, 1960.
13. Luc Boltanski and Eve Chiappello, *The New Spirit of Capitalism*, London: Verso, 2004 [1999].
14. Manuel Castells, *The Internet Galaxy: Reflections on the Internet, Business and Society*, Oxford and New York: Oxford University Press, 2001; Thomas Malone, *The Future of Work: How the New Order of Business Will Shape Your Organisation, Your Management Style and Your Life*, Boston, MA: Harvard Business School Press, 2004.
15. Charles Hecksher and Anne Donnellon, *The Post-Bureaucratic Organisation: New Perspectives on Organisational Change*, London: Sage, 1994.
16. David Osborne and Ted Gaebler, *Re-Inventing Government: How the Entrepreneurial Spirit Is Transforming the Public Sector*, Reading, MA: Addison-Wesley, 1992.
17. Paul du Gay, 'Against "Enterprise" (but not against "enterprise", for that would make no sense)', *Organization*, Vol. 11, No, 1 (2004), pp. 37–57.
18. Manuel Castells, *The Rise of the Network Society* (Vol. 1 of *The Information Age: Economy, Society and Culture*), 2nd edn, London: Blackwell, 2000, p. 500.
19. Manuel Castells, *The Power of Identity* (Vol. 2 of *The Information Age: Economy, Society and Culture*), 2nd edn, London: Blackwell, 2004.
20. Barbara Epstein, 'Anarchism and the anti-globalisation movement', *Monthly Review*, Vol. 53, No. 4 (September 2001) [online].
21. Jannis Kallinikos, 'ICT, organisations and networks', in Robin Mansell, Chrisanthi Avgerou, Danny Quah and Roger Silverstone (eds), *Oxford Handbook of Information and Communication Technologies*, Oxford: Oxford University Press, 2007, p. 286.
22. Kallinikos, 'The social foundations', p. 24.
23. James G. March and Johan P. Olsen, *Ambiguity and Choice in Organisations*, Oslo: Universitetsfoerlaget, 1976.
24. James R. Barker, *The Discipline of Teamwork: Participation and Concertive Control*, Thousand Oaks, CA. Sage; David Courpasson, 'Managerial strategies of domination: power in soft bureaucracies', *Organisation Studies*, Vol. 21, No. 1 (2000), pp. 141–61.
25. Kallinikos, 'The social foundations', p. 30.

26. Andrea Forte and Amy Bruckman, 'Scaling consensus: increasing decentralisation in Wikipedia governance', in *Proceedings of the Hawaiian International Conference of Systems Sciences*, January 2008, p. 157.

27. Yochai Benkler, *The Wealth of Network: How Social Production Transforms Markets and Freedom*, New Haven, CT: Yale University Press, 2006.

28. James Coleman, 'Authority systems', *Public Opinion Quarterly*, Vol. 44, No. 2 (Summer 1980), p. 150.

29. Ibid, p. 158. An illustration of this point is the acronym 'NSFW' found in blogs next to content (such as embedded videos) which is deemed 'not safe for work'.

30. Emily Erikson and Joseph M. Parent, 'Central authority and order', *Sociological Theory*, Vol. 25, No. 3 (2008), p. 250.

31. See Joyce Rothschild and Raymond Russell, 'Alternatives to bureaucracy: democratic participation in the economy', *Annual Review of Sociology*, Vol. 12 (1986), pp. 307–28.

32. Rothschild-Whitt, 'The collectivist organisation'.

33. Ibid.

34. Coleman, 'Authority systems', p. 158.

35. Mark A. Lemley 'The law and economics of Internet norms', *Chicago–Kent Law Review*, Vol. 73 (1997), pp. 1267–70.

36. Ibid, p. 1268.

37. Jane Siegel, Vitaly Dubrovsky, Sara Kiesler and Timothy W. McGuire, 'Group processes in computer-mediated communication', *Organisational Behaviour and Human Decision Processes*, No. 37 (1986), pp. 157–87.

38. Jack Glaser and Kimberly Kahn, 'Prejudice, discrimination and the Internet', in Yair Amichai-Hamburger (ed.), *The Social Net: Understanding Human Behaviour in Cyberspace*, Oxford and New York: Oxford University Press, 2005, p. 261; Katelyn Y.A. McKenna and John Bargh, 'Plan 9 from cyberspace: the implications of the Internet for personality and social psychology', *Personality and Social Psychology Review*, No. 4 (2000), pp. 57–75.

39. Luc Boltanski and Laurent Thévenot, *On Justification: Economies of Worth*, Princeton, NJ: Princeton University Press, 2006 [1991].

40. Pamela J. Hinds and Diane E. Bailey, 'Out of sight, out of sync: understanding conflict in distributed teams', *Organisation Science*, Vol. 14, No. 6 (November–December 2003), p. 618.

41. Ibid, p. 623.

42. Lisa Hope Pelled, Kathleen M. Eisenhardt and Katherine R. Xin, 'Exploring the black box: an analysis of work group diversity,

conflict, and performance', *Administrative Science Quarterly*, Vol. 44, No. 1 (1999), pp. 1–28.

43. Karen A. Jehn, 'A quantitative analysis of conflict types and dimensions in organisational groups', *Administrative Science Quarterly*, Vol. 42, No. 3 (1997), pp. 530–58.

44. Elizabeth A. Mannix, Terri Griffith and Margaret A. Neale, 'The phenomenology of conflict in distributed work teams', in Pamela J. Hinds and Sara Kiesler (eds), *Distributed Work*, Cambridge MA: MIT Press, 2002, pp. 213–33.

45. Caprice may legitimise charismatic authority, since being systematic means one can be replaced. See Erikson and Parent, 'Central authority and order', p. 256.

46. Vivian Franco, Hsiao-Yun Hu, Bruce V. Lewenstein, Roherta Piirto, Russ Underwood and Noni Korf Vidal, 'Anatomy of a flame: conflict and community building on the Internet', in Eric Lesser, Michael Fontaine and Jason Slusher (eds), *Knowledge and Communities*, Newton, MA: Butterworth–Heinemann, 2000, pp. 209–224.

47. Jeffrey S. Juris, 'Networked social movements: global movements for global justice', in Manuel Castells (ed.), *The Network Society: A Cross-Cultural Perspective*, Cheltenham: Edward Elgar, p. 357.

48. Michel Maffesoli, *The Time of the Tribes: The Decline of Individualism in Mass Societies*, London, Thousand Oaks, CA and New Delhi: Sage, 1996.

49. Stephen Coleman, 'E-Democracy: the history and future of an idea', in Mansell et al., *Oxford Handbook*, p. 370.

50. Henry Perritt, 'Cyberspace self-government: town-hall democracy or rediscovered royalism?', *Berkeley Technology Law Journal*, Vol. 12 (1997), p. 475.

51. Jaron Lanier, 'Digital Maoism: the hazards of the new online collectivism', *The Edge*, May 2006 [online].

52. Auray, 'Le Modèle souverainiste'.

53. These remarks only apply to people in liberal democracies. Citizens of Iran or China, for example, may face imprisonment or ill-treatment because of their online activities.

54. Lee Salter, 'Democracy, new social movements and the Internet: a Habermasian analysis', in Martha McCaughey and Michael D. Ayers (eds), *Cyberactivism: Online Activism in Theory and Practice*, New York and London: Routledge, 2003, pp. 117–44.

55. Christian Fuchs, *Internet and Society: Social Theory in the Information Age*, London and New York: Routledge, 2008, p. 164.

56. Rothschild-Whitt, 'The collectivist organisation', p. 522.

57. Ibid.
58. Robert Castel, 'La critique face au marché', in Jean Lojkine (ed.), *Les Sociologies critiques du capitalisme*, Paris: Presses Universitaires de France, 2002, pp. 207–13.
59. Ibid.
60. Max Weber, *Economy and Society: An Outline of Interpretive Sociology*, Berkeley, Los Angeles and London: University of California Press, 1978 [1922].
61. See Lawrence H. Keeley, *War Before Civilisation: The Myth of the Peaceful Savage*, New York: Oxford University Press, 1996.

INDEX

Compiled by Sue Carlton